Edited by
Manfred Weidmann, Nigel Silman, Patrick Butaye, and Mandy Elschner

Working in Biosafety Level 3 and 4 Laboratories

Related Titles

Elschner, M., Cutler, S., Weidmann, M., Butaye, P. (eds.)

BSL3 and BSL4 Agents

Epidemiology, Microbiology, and Practical Guidelines

2012
Print ISBN: 978-3-527-31715-8,
also available in digital formats

Bergman, N.H.

Bacillus anthracis and Anthrax

2011
Print ISBN: 978-0-470-41011-0,
also available in digital formats

Katz, R., Zilinskas, R.A. (eds.)

Encyclopedia of Bioterrorism Defense, 2nd Edition

2 Edition
2011
Print ISBN: 978-0-470-50893-0,
also available in digital formats

Kostic, T., Butaye, P., Schrenzel, J. (eds.)

Detection of Highly Dangerous Pathogens

Microarray Methods for BSL3 and BSL4 Agents

2009
Print ISBN: 978-3-527-32275-6,
also available in digital formats

Stulik, J., Toman, R., Butaye, P., Ulrich, R.G. (eds.)

BSL3 and BSL4 Agents

Proteomics, Glycomics, and Antigenicity

2011
Print ISBN: 978-3-527-32780-5,
also available in digital formats

Edited by Manfred Weidmann, Nigel Silman, Patrick Butaye and Mandy Elschner

Working in Biosafety Level 3 and 4 Laboratories

A Practical Introduction

WILEY Blackwell

Editors

Dr. Manfred Weidmann
University Medical Center Göttingen
Department of Virology
Kreuzbergring 57
37075 Göttingen
Germany

Prof. Nigel Silman
Research & Development
Public Health England
Porton Down
Salisbury SP4 0JG
United Kingdom

Prof. Patrick Butaye
Veterinary and Agrochemical Research Centre
Department of Bacteriology and Immunology
Groeselenberg 99
1180 Bruxelles
Belgium

Dr. Mandy Elschner
Friedrich-Loeffler-Institute
Federal Research Institute for Animal Health
Institute of Bacterial Infections and Zoonoses
Naumburger Straße 96 a
07743 Jena
Germany

Cover

Working at a biosafety cabinet class II wearing a positive respirator (Photo: Martin Spiegel, University of Göttingen)

1st Reprint 2014

Library of Congress Card No.: applied for

British Library Cataloguing-in-Publication Data
A catalogue record for this book is available from the British Library.

Bibliographic information published by the Deutsche Nationalbibliothek
The Deutsche Nationalbibliothek lists this publication in the Deutsche Nationalbibliografie; detailed bibliographic data are available on the Internet at <http://dnb.d-nb.de>.

Wiley-Blackwell is an imprint of John Wiley & Sons, formed by the merger of Wiley's global Scientific, Technical, and Medical business with Blackwell Publishing.

Print ISBN: 978-3-527-33467-4
ePDF ISBN: 978-3-527-67536-4
ePub ISBN: 978-3-527-67533-3
Mobi ISBN: 978-3-527-67534-0
oBook ISBN: 978-3-527-67535-7

Cover Design Graphik Design Schulz, Fußgönheim, Germany
Typesetting Laserwords Private Limited, Chennai, India
Printing and Binding Markono Print Media Pte Ltd, Singapore

Printed on acid-free paper

Contents

Acknowledgement

This project has been funded with support from the European Commission. This publication reflects the views only of the author, and the European Commission cannot be held responsible for any use which may be made of the information contained therein.

With the financial support of the Prevention of and Fight against Crime Programme European Commission – Directorate-General Home Affairs

Preface

This is the fourth book written in a series that started within the framework of the European project COST (European Cooperation in Science and Technology) action B28, which aimed at increasing knowledge on BSL-3 (biosafety level) and BSL-4 agents; to support development of more accurate diagnostic assays, vaccines, and therapeutics; and to better understand epidemiology of these highly pathogenic microorganisms that can potentially be used as biological weapons. COST funding has ended in 2010 and an additional grant by DG Home Affairs in the program "Prevention of and fight against crime" now sponsors the final volume "A practical guideline to working in BSL-3/4 laboratories."

The first book summarized the knowledge on microarray technology. The second book summarized the knowledge on proteomics, glycomics, and antigenicity of the BSL-3/4 agents. The third book described the agents themselves.

The authors of the chapters are all involved in research concerning these agents and have been working with them extensively. Typically, the authors also have access to BSL-3/4 laboratories in which they can work with these agents, thus are familiar with the required precautions and legislations. Their expertise has also been employed for assessment of outbreaks and understanding the epidemiological factors that facilitate their spread and subsequent control.

This book was created from lectures given in a training school course held four times funded by COST action from 2007 to 2010 and four times funded by DG Home in 2012–2013. At the time of publication, two more courses have to be held.

Manfred Weidmann
Nigel Silman
Patrick Butaye
Mandy Elschner

List of Contributors

Francesc Xavier Abad
CReSA
Centre de Recerca en Sanitat
Animal
UAB-IRTA
Bellaterra
08193 Barcelona
Spain

Åsa S. Björndal
Swedish Institute for
Communicable Disease Control
Unit for Biorisk Management
Department för Analysis and
Prevention
Nobels väg 18
SE-17182 Solna
Sweden

Patrick Butaye
Veterinary and Agrochemical
Research Centre
Department of Bacteriology and
Immunology
Groeselenberg 99
1180 Brussels
Belgium

Mariano Domingo
CReSA
Centre de Recerca en Sanitat
Animal
UAB-IRTA
Bellaterra
08193 Barcelona
Spain

Mandy Elschner
Friedrich-Loeffler-Institut
Federal Research Institute for
Animal Health
Institute of Bacterial Infections
and Zoonoses
Naumburger Straße 96a
07743 Jena
Germany

Martin Heller
Friedrich-Loeffler-Institut
Federal Research Institute for
Animal Health
Institute of Molecular
Pathogenesis
Naumburger Straße 96a
07743 Jena
Germany

Frank T. Hufert
University Medical Center
Göttingen
Department of Virology
Kreuzbergring 57
37075 Göttingen
Germany

Annette A. Kraus
Swedish Institute for
Communicable Disease Control
(SMI)
Department for Diagnostics and
Vaccinology
Nobels väg 18
SE-17182 Solna
Sweden

Jürgen Mertsching
Hannover Medical School
Department of Biological Safety
Carl-Neuberg-Straße 1
30625 Hannover
Germany

Ali Mirazimi
Swedish Institute for
Communicable Disease Control
(SMI)
Department for Diagnostics and
Vaccinology
Nobels väg 18
SE-171 82 Solna
Sweden

Nigel Silman
Research & Development
Public Health England
Porton Down
SP4 0JG Salisbury
United Kingdom

David Solanes
CReSA
Centre de Recerca en Sanitat
Animal
UAB-IRTA
Bellaterra
08193 Barcelona
Spain

Manfred Weidmann
University Medical Center
Göttingen
Department of Virology
Kreuzbergring 57
37075 Göttingen
Germany

Introduction

Manfred Weidmann

Biosecurity practices cannot be built without a strong safety culture; therefore, it is generally agreed that training for BSL-3/4 (biosafety level) work should be a precondition for starting to work in these specialized, safety and security sensitive, laboratories. Training for work in BSL-3/4 laboratories, however, is hardly available on an organized level and mainly performed in a traditional laboratory culture of individual training. In recent years, however, the demand for broader training of ever more young scientists working in newly created BSL-3/4 laboratories has increased (at the latest count, there were over 600 BSL-3 laboratories in the United States) [1]. The steady rise of people working in this type of laboratory environment may inflate the potential for accidents, as has been seen in the United States in recent years. Therefore, organized training, as an addition to the individual training, is absolutely necessary and indeed has been announced as obligatory in the United States [2]. In Europe, there are several projects offering training, for example, Euronet-P4 has provided additional practical training for BSL-4 laboratory staff, ETIDE has provided training for clinicians for infectious disease emergencies since 2007, and Biosafety Europe has provided safety guidelines [3].

Participants of the COST Action B28 "Array technologies for BSL-3 and BSL-4 pathogens" [4] developed a new training course for scientists working in BSL-3/4 laboratories and ran it once a year from 2007 to 2010. Funding by DG Home Affairs allows for two courses a year since 2012 [5].

The 4-day course consists of lectures and practical training. The background of the lecturers represents all the very different possibilities of organizing a BSL-3/4 laboratory including the adaptation to the local requirements of biosafety, safety at work, and social regulations.

Working in a BSL-3 laboratory is a very good basis to commence work at BSL-4 level, especially, as the danger of infection in a BSL-3 level (a biosafety shell to protect the environment) can be considered to be potentially much higher than that in a BSL-4 laboratory, which offers superior personal protection. The BSL-3 facilities at the Department of Virology, Göttingen University Medicine allow the simultaneous training of 10 students, two each working in one class-2 biological safety cabinet supervised by one lecturer. As training in a BSL-4 at course level

Working in Biosafety Level 3 and 4 Laboratories: A Practical Introduction, First Edition.
Edited by Manfred Weidmann, Nigel Silman, Patrick Butaye, and Mandy Elschner.

of this size is impossible due to security and infrastructure reasons, this course provides at least a good basic practical introduction to the principles of working at BSL-3 that are also adhered to in a BSL-4 environment.

The lectures summarized in this book cover biocontainment, hazard criteria and categorization of microbes, technical specifications of BSL-3 and ABSL-3 (animal biosafety level) laboratories, personal protective gear, shipping BSL-3 and BSL-4 organisms according to UN and IATA (International Air Transport Association) regulations, efficacy of inactivation procedures, fumigation, learning from a history of laboratory accidents, handling samples that arrive for diagnostic testing, and bridging the gap between the requirements of biocontainment and diagnostics. The practical sessions of this course are run over three afternoons and cover the use of personal safety equipment including the use of positive pressure masks, dexterity training, and inactivation procedures for viruses and bacteria [6]. Altogether, the combination of lectures and practicals provides a good focused introduction into principles and regulations important for both types of laboratory environments.

A general conflict between laboratory scientists and biosafety officers in all countries appears to be the difficulty of convincing biosafety officers to accept proof of biosafety that has not yet been cast into official rules and regulations. On the other hand, it is quite clear that regulations cannot cover all aspects of working in these laboratories especially when dealing with new organisms. The SARS-CoV (severe acute respiratory syndrome coronavirus) outbreak in 2003 can be seen as a showcase for this dilemma [7].

Biosafety regulations issued are often not flexible enough and can tend to impede work rather than to increase safety. The notion that the scientists at work cannot be trusted seems contradictory to the fact that the scientists, working with highly contagious and pathogenic agents, have an eminent interest in their own safety and health.

A major effort across the European Union therefore should be to build a flexible framework to accept biosafety evidence created locally by the different respective laboratories. There is a need for a consensus between biosafety organizations and the scientists at work, on how proof of biosafety should be shown and documented in order to sustain and not stifle a flexible capacity to deal with novel and well-known pathogens.

Tighter regulations issued by national biosafety bodies in response to the green paper launched by the European Commission [8] may blur or bias their perception of biosafety necessities, which might make working in BSL-3/4 laboratories close to impossible. It would be ironic if a measure initiated because of political biothreat considerations would essentially impede all biopreparedness actions toward unexpected events in the public domain.

Recently, Kimman *et al.* [7] have provided a literature review on laboratory-acquired infections (LAIs) including those that occurred in the wake of the SARS-CoV outbreak. Their conclusion was that deviation from general "good microbiological practice" is the most frequent cause for LAI and that training for compliance to procedures and regulations appears to be the best method to avoid these. In this, the lectures summarized here offer the opportunity to improve the available basic training for BSL-3/4 scientists and prerequisite initial training for beginners.

References

1. Gronvall, G.K., Fitzgerald, J., Chamberlain, A., Inglesby, T.V., and O'Toole, T. (2007) High-containment biodefense research laboratories: meeting report and center recommendations. *Biosecur. Bioterror.*, **5**, 75–85.
2. Le Duc, J.W., Anderson, K., Bloom, M.E. *et al.* (2008) Framework for leadership and training of biosafety level 4 laboratory workers. *Emerging Infect. Dis.*, **14**, 1685–88.
3. Ippolito, G., Nisii, C., and Capobianchi, M.R. (2008) Networking for infectious-disease emergencies in Europe. *Nat. Rev. Microbiol.*, **6**, 564.
4. COST B28 *http://www.cost-b28.be/index.php/pages/index* (accessed 13 April 2013).
5. Weidmann, M., Hufert, F., Elschner, M. *et al.* (2009) Networking for BSL-3/4 laboratory scientist training. *Nat. Rev. Microbiol.*, **7**, 756.
6. Abteilung Virologie *http://www.virologie.uni-goettingen.de/index.php?page=22&empty=1&id=26* (accessed 13 April 2013).
7. Kimman, T.G., Smit, E., and Klein, M.R. (2008) Evidence-based biosafety: a review of the principles and effectiveness of microbiological containment measures. *Clin. Microbiol. Rev.*, **21**, 403–25.
8. European Commission (2007) Green Paper on Bio-Preparedness.

1
Laboratory Biosafety in Containment Laboratories

Annette A. Kraus and Ali Mirazimi

Microbiological research on and diagnostics of highly pathogenic microorganisms, namely bacteria and viruses, have to be conducted in containment laboratories in order to contain the infectious material. We are therefore referring to the "Concept of Biocontainment" – a concept, which dates back to the 1940s, when the first biosafety cabinet (BSC) class III was developed at the US Army Biological Warfare Laboratories in Fort Detrick, Maryland.

Biocontainment is required to prevent accidental infection of researchers or diagnostic staff and to avoid release of the infectious agents into the surrounding environment.

1.1
Routes of Infection

Depending on the nature of the microorganism, there is a great variety of infection routes. However, the natural route of transmission may be different in a laboratory setting when working with isolated pathogens. This has to be considered when establishing working procedures for biosafety laboratories.

In a research laboratory, for instance, HIV or hepatitis B virus is not transmitted via the natural route, that is, from person to person through direct contact of body fluids, but, for example, through accidental inoculation with a syringe. The bacterial pathogen *Neisseria gonorrhoeae* spreads through direct contact under normal circumstances and laboratory workers have also to primarily protect themselves from direct contact (Chapter 10).

Bacteria and viruses that are vector-borne such as the tick-borne encephalitis virus (TBE) or the tick-borne *Borrelia* sp. obviously have a different infection route in the laboratory as compared to the natural setting. Here, protection should also aim at preventing needlestick injuries with contaminated syringes and direct contact with fluids that have a high concentration of the infectious agent.

While some other bacteria such as *Salmonella* or *Vibrio cholerae* spread via the fecal-oral route through contaminated food or water and are relatively easy to contain, others spread readily via aerosols and are more difficult to control.

Working in Biosafety Level 3 and 4 Laboratories: A Practical Introduction, First Edition.
Edited by Manfred Weidmann, Nigel Silman, Patrick Butaye, and Mandy Elschner.

Consequently, when working with bacteria such as *Mycobacterium tuberculosis*, laboratory workers have to protect themselves by wearing personal protection devices (respirators) and perform the work in BSCs. The same measures have to be taken when working with viral pathogens that also spread through aerosols, such as the avian influenza virus or hantaviruses.

In general, airborne or aerosol-transmitted pathogens are comparatively difficult to work with. Aerosols are practically invisible to the human eye, not to mention airborne viruses or bacteria. Using syringes, pipettes, and mixing devices, even according to good laboratory practice (GLP) protocols, creates aerosols in unexpected amounts. Dimmick *et al.* conducted a study in the early 1970s and estimated the aerosol dose originating from pipetting 1 ml to be 10^{10} small particles. Even at a distance of 3 m, the number of small particles still reaches 50. Depending on the nature of the pathogen, this can clearly reach or exceed the infectious dose of that particular virus or bacterium.

Generally, different precautionary measures have to be applied when working with different pathogens according to the routes of transmission.

1.2 Classification of Microorganisms

When establishing routines for microbiological laboratories, not only the route of infection of the used pathogens has to be considered, but the infectious dose, available countermeasures, and preexisting immunity have to be considered as well. In addition, information about concentration of the isolated pathogen, total volumes used in a certain research or diagnostic setting, as well as experience is important when defining the risk.

While airborne viruses belong to the most difficult pathogens to contain, work with varicella virus, for instance, can be performed under moderate safety precautions, because the disease is treatable and a vaccine is available. HIV, on the other hand, causes a lethal disease that can be neither treated nor prevented by a vaccine, is a rather unstable virus with a comparatively low infectious dose, and can therefore also be handled with moderate safety precautions.

The obvious question is now to classify the microorganisms and translate this classification into levels of precautionary measures, that is, into biological safety levels (BSLs). The levels of containment range from the lowest safety level 1 to the highest at level 4. In the United States, the Centers for Disease Control and Prevention (CDC) have specified these levels. In the European Union, the same BSLs are defined in a directive (Commission Directive 97/65/EC). In summary, the classification of microorganisms is based on various parameters specific for every pathogen including routes of transmission, severity of the disease, infectious dose, available countermeasures or preventive measures, and transmissibility to the community. These may be influenced by existing levels of immunity, density and movement of host population presence of appropriate vectors, and standards of environmental hygiene.

The WHO (World Health Organization) has classified infectious microorganisms by risk groups and the following list provides an overview of the risk levels when working with different pathogens. There are however other classification schemes (Chapter 2):

WHO risk group 1 – minimal risk
Microorganisms that usually do not cause human disease, such as *Escherichia coli* K12 or *Lactobacillus*.

WHO risk group 2 – moderate risk
Microorganisms that cause treatable or self-healing diseases and are difficult to contract via aerosol in a laboratory setting, such as salmonella or measles virus.

WHO risk group 3 – high risk
Highly contagious microorganisms that cause serious diseases, such as TBE virus or *M. tuberculosis*.

WHO risk group 4 – very high risk
Highly contagious microorganisms that cause serious diseases, even epidemics, with high mortality rate, such as Ebola virus or Lassa fever virus.

1.3 General Containment Principles

In general, there are two different levels of protection against accidental infections when working with pathogens in a research or diagnostic laboratory, a primary and a secondary barrier. In addition to these, safe working procedures and techniques together with safety equipment complement a containment laboratory.

Primary containment provides the protection of personnel and the immediate laboratory environment from exposure to infectious agents and is provided by good microbiological technique and the use of appropriate safety equipment, such as BSC. Secondary containment is the protection of the environment external to the laboratory from exposure to infectious materials and is provided by a combination of facility design and operational practices.

1.4 Specific Containment Principles

At the lowest level of biocontainment, the containment zone may only be a chemical fume hood. At the highest level, the containment involves isolation of the organism by means of building systems, sealed rooms, sealed containers, personal protective equipment, and detailed procedures for entering the laboratories, coupled with decontamination procedures when leaving them. In most cases, this also includes high levels of security for access to the facility, ensuring that only authorized personnel may be admitted to such laboratories.

The following list describes the different specific measures of the BSLs 1–4 laboratories.

1.4.1 Biosafety Level 1 Laboratory

This level is suitable for work involving well-characterized agents not known to consistently cause disease in healthy adults and of minimal potential hazard to laboratory personnel and the environment (CDC, 1997). At this level, precautions against the biohazardous materials in question are minimal, most likely involving gloves and perhaps some sort of facial protection (if indicated by risk assessment). Decontamination procedures for this level are similar in most respects to modern precautions against everyday microorganisms, that is, washing hands with antibacterial soap and/or washing all exposed surfaces of the laboratory with disinfectants.

1.4.2 Biosafety Level 2 Laboratory

This level is similar to BSL-1 and is suitable for work involving agents of moderate potential hazard to personnel and the environment. Here, laboratory staff have specific training in handling pathogenic agents, and consequently, access to the laboratory is restricted. Extreme precautions are taken with contaminated syringes and other sharp items to avoid accidental infections. In addition, certain procedures in which infectious aerosols or splashes may be created are conducted in a BSC.

1.4.3 Biosafety Level 3 Laboratory

This level is applicable to research and diagnostic laboratories in which work is done with indigenous or exotic agents, which may cause serious or potentially lethal disease on transmission. Laboratory staff have specific training in handling pathogenic agents. All procedures involving the manipulation of infectious materials are conducted within a BSC, or other physical containment devices (engineering controls), and by personnel wearing appropriate personal protective clothing and equipment. The laboratory has special engineering and design features, so that, for instance, the filtered exhaust air from the laboratory room is discharged to the outside. In addition, the laboratory is negatively pressurized and entry and exit are therefore limited through an air lock. Access to the laboratory is restricted when work is in progress, and the recommended standard microbiological practices and safety procedures for BSL-3 are rigorously followed. Basically however a BSL-3 laboratory is designed to protect the environment from contamination by pathogens.

It does not increase safety for laboratory staff; this is the function of the primary engineering controls employed.

1.4.4 Biosafety Level 4 Laboratory

Dealing with biological hazards at this level requires either suit-based or cabinet-line-based BSL-4 laboratories. As there are only a few biocontainment laboratories that use class III biological safety cabinets, this chapter focuses on the use of a one-piece positive pressure personnel suit in combination with a class II biological safety cabinet and a self-contained breathing-air supply in a high-containment laboratory with negative pressure.

1.5 Design of a Suit-Based-BSL-4 Laboratory with Negative Pressure

Building a BSL-4 laboratory is a huge technical undertaking. For every square meter of laboratory space, 5 m^2 of technical area is required. In principle, all that is needed is a room with a BSC. This room however has to be completely isolated from other rooms in the building, and ideally, the facility is either in a separate building or in a controlled area, which is completely isolated from all other areas of the building.

Rooms in the facility must be arranged to ensure exit by sequential passage through the chemical shower, inner (dirty) suit room, personal shower, and outer (clean) changing area. An air lock fitted with airtight doors is used during personnel entry and exit. To enter and move material to and from the laboratory, electronically secured air locks are employed to prevent both doors being opened at the same time.

Safe entry and exit procedures have to be established for the laboratory staff. The chemical shower is used to decontaminate the surface of the positive pressure suit before the worker leaves the laboratory. In addition, there must be a system to take materials of various natures, ranging from heavy technical equipment to highly sensitive samples, into the laboratory and potentially out again. This could be a fumigation chamber or a dunk tank filled with a decontaminant.

The BSL-4 laboratory as a sealed entity has special engineering and design features to prevent microorganisms from being disseminated into the environment.

To this aim, the laboratory is kept at negative air pressure, so that air flows into the room if the barrier is penetrated or breached. The supply and exhaust components of the filtered ventilation system must be designed to guarantee a set rate of air changes, to maintain the negative pressure to surrounding areas, and to provide differential pressure or directional airflow as appropriate between adjacent areas within the laboratory. Several concepts of BSL-4 laboratory structure have been designed. Sandwich type laboratories house the laboratory floor in between technical floors generally providing supplies from above and treating

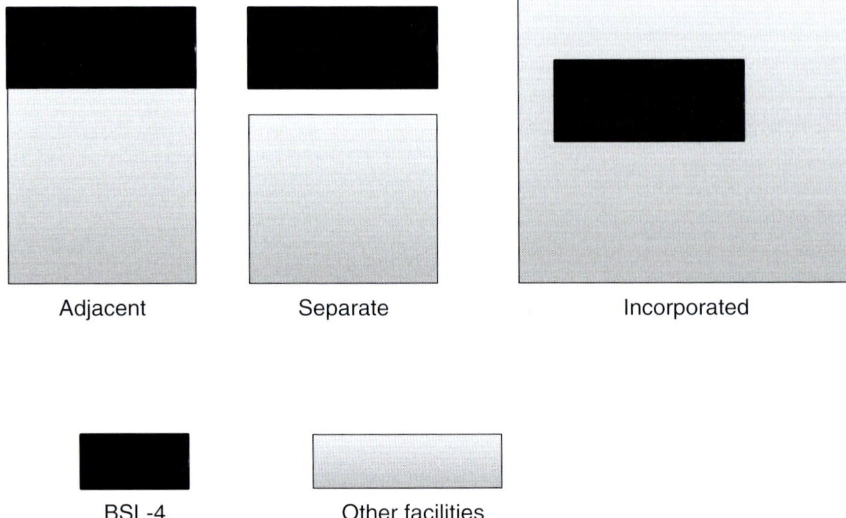

Figure 1.1 BSL-4 suit models.

waste and effluents below. The technical and laboratory sections can however also be arranged adjacent to each other or the laboratory space is arranged inside the shell of the technical section in a box-in-a-box containment approach [1] (Figure 1.1).

All work in the level 4 laboratory is done in a pressurized and ventilated suit. Air for breathing is passed into the suit through a hose and is filtered to be free of microorganisms. In addition, most activities involving pathogens in the work areas of the facility are confined to a class II BSC. In contrast to a BSL-3 laboratory, a BSL-4 laboratory therefore provides maximum protection for laboratory staff as well as preventing release of pathogens to the environment.

To perform laboratory work, access to water and obviously to a sewer system is necessary. All air and water services coming from a BSL-4 laboratory will undergo decontamination procedures to eliminate the possibility of an accidental release. Waste will be decontaminated through double-door autoclaves.

Walls, floors, and ceilings of the laboratory must be constructed to form a sealed internal shell to facilitate fumigation and prohibit animal and insect intrusion. The internal surfaces of this shell must be resistant to chemicals used for cleaning and decontamination of the area.

For safety as well as practical reasons, communication devices have to be installed providing telephone, fax, and data lines. Research and diagnostic laboratory members as well as technical staff have to be trained in handling hazardous infectious agents. They must understand the primary and secondary containment functions of the standard and special practices, the containment equipment, and the laboratory design characteristics. Finally, all work done in the laboratory needs to be documented and access to the laboratory is strictly controlled.

1.6 Safety Routines

In order to maintain a high level of safety as well as security in a biosafety laboratory, certain rules and routines have to be established and strictly enforced. Even though the regulations will vary to a certain extent in different laboratories – being situated in different countries with specific directives, handling particular specimens or based on specific physical premises – the regulations are in principle the same.

Primarily, access to the containment laboratory is restricted to authorized personnel only (who has passed a designed course for high containment laboratory level 4). Research and diagnostic staff are thoroughly checked and trained before being granted access to the laboratory. The training includes several working hours under supervision of a mentor with experience in the biosafety laboratory. Annual exercises of the safety routines are mandatory and ensure a high standard of preparedness for the case of an emergency.

The daily routines for working in the BSL-4 laboratory are based on a "buddy system" or camera-based system and include a list of checkpoints. There should also be a decontamination team available outside of the laboratory, which provides trained help in case of an emergency.

Before the BSL-4 laboratory can be entered and work can commence, a list of key points has to be checked by the research or diagnostic personnel. A number of technical parameters will be controlled every time, such as the pressure status of the laboratory and the adjacent rooms, or whether the internal communication system is working. In addition, the filling status of the chemical shower supply tanks is checked as well as the air supply and ventilation systems. The actual access to the laboratory is then restricted to authorized staff. Once inside the laboratory wing and changing rooms, researchers change into common laboratory clothing and activate the personal communication devices. The overpressure safety suits will be put on and connected to the air supply in the so-called suit room. Finally, the researchers gain access to the actual work place through an air lock. A recommended working period of 3–5 h should not be exceeded. Exit from the laboratory is essentially the reverse process with the addition of a suit decontamination step in the chemical shower.

Summary

Outbreaks of emerging infectious diseases continue to challenge both human and veterinary health in Europe and around the world. Events such as the outbreaks of SARS-CoV (severe acute respiratory syndrome coronavirus), H5N1 avian influenza, Ebola virus in the Congo, Lassa fever in West Africa, and Crimean–Congo hemorrhagic fever virus (CCHFV) in Europe. To develop strategies to prevent outbreaks and combat these diseases, we need to develop research and diagnostic platforms. Research and diagnostics of highly pathogenic microorganism have to be conducted in containment laboratories in order to contain the infectious material. The WHO has classified infectious microorganisms by risk groups from 1 to 4. In this chapter, we discuss the biocontainment that can be used to handle risk class IV pathogens. Dealing with biological hazards at this level requires either suit-based or cabinet-line-based BSL-4 laboratories. In this chapter, we focus on the use of a one-piece positive pressure personnel suit in combination with a class II biological safety cabinet and a self-contained breathing-air supply in a high-containment laboratory with negative pressure.

Reference

1. Crane, C.T., Bullock, C.F., and Richmond, J.Y. (1999) Designing the BSL4 laboratory (chapter 9). *J. Am. Biol. Saf. Assoc.*, **4** (1), 24–32.

2
Hazard Criteria and Categorization of Microbes Classification Systems

Nigel Silman

Microorganisms are classified into different groups according to the risk that they pose to laboratory workers, the general public, and the environment, if they should be released from the laboratory. Each country in the world has its own system of classification; therefore, the same microorganisms may be differently categorized depending on local regulations. In general, risk-based approaches are used, which are based on the World Health Organization's (WHO) definitions; the risk assessment that must be made considers the pathogenicity of the infectious agent, its mode of transmission, and whether or not there are insect vectors or animal hosts; the availability of prophylactic and postexposure therapeutics or vaccines; the likelihood of spread to the community; and the availability of control measures, for example, food and water hygiene measures. These recommendations stem from the 1970s when it was realized that some pathogenic microorganisms were more likely to cause laboratory acquired infections in workers handling them than others and that laboratories should be designed and working practices enforced that are proportionate to the risk the pathogen poses to the laboratory worker as well as the environment and general public.

The *biological agent* may be defined as a microorganism, cell culture, or human endoparasite, whether or not genetically modified, which may cause infection, allergy, toxicity, or otherwise create a hazard to human health. In considering categorization of biological agents, this definition will be used. All classification schemes currently in operation are based on four categories of pathogenic microorganisms; these categories may be based on the propensity of the microorganism to cause human or animal disease. The definitions of the four categories (as defined by the WHO) are as follows:

Hazard group 1. A biological agent that is unlikely to cause human disease;

Hazard group 2. A pathogen that can cause human or animal disease but is unlikely to be a serious hazard to laboratory workers, the community, livestock, or the environment. Laboratory exposures may cause serious infection, but effective treatment and preventative measures are available and the risk of spread of infection is limited.

Working in Biosafety Level 3 and 4 Laboratories: A Practical Introduction, First Edition.
Edited by Manfred Weidmann, Nigel Silman, Patrick Butaye, and Mandy Elschner.

In contrast, the health and safety executive in the United Kingdom acting through the control of substances hazardous to health (COSHH) regulations defines the same group of microorganisms as follows:

A biological agent that can cause human disease and may be a hazard to employees; it is unlikely to spread to the community and there is usually effective prophylaxis or treatment available.

The difference between these two definitions resides in the introduction of animal disease in the WHO definition, whereas COSHH is a set of regulations designed to prevent risks to human health; thus, animal infection is not mentioned and the emphasis is about the biological agent that may have the potential to cause harm to humans. The WHO definition however states that these microorganisms are "unlikely to be a serious hazard to laboratory workers." The point of comparing these two definitions is that when making risk-based assessments of pathogenic microorganisms, the risk may be applied not only to the laboratory worker but also to animals of economic importance and also cover the risk of transmission to the community if an accidental release should occur from the laboratory.

Hazard group 3. A pathogen that usually causes serious human or animal disease but does not ordinarily spread from one infected individual to another. Effective treatment and preventative measures are available.

For comparison, here is the COSHH hazard group 3 (HG-3) definition, again the emphasis is on harm to the laboratory worker:

A biological agent that can cause severe human disease and may be a serious hazard to employees: it may spread to the community, but there is usually effective prophylaxis or treatment available.

Of course, both definitions are trying to cover the risk to the laboratory worker primarily, but HG-3 microorganisms are often described as those that pose high individual risk but low community risk. Bear in mind that this is a generalization and there are specific examples of HG-3 organisms that pose very particular risks to the community; examples of these are highly pathogenic avian influenza viruses and SARS (severe acute respiratory syndrome) viruses, both of which are capable of sustained community transmission if they should be accidentally released, and on this basis do not fit the definition of a HG-3 organism particularly well. A bacterial example would be *Yersinia pestis* the causative agent of plague; once again, pneumonic plague is a highly transmissible disease to the community although bubonic plague requires flea vectors and rat hosts for sustained community transmission.

Hazard group 4. A pathogen that usually causes serious human or animal disease and that can be readily transmitted from one individual to another, directly or indirectly. Effective treatment and preventative measures are not usually available.

Again, for comparison, COSHH defines HG-4 microorganisms as follows:

A biological agent that can cause severe human disease and is a serious hazard to employees; it is likely to spread to the community and there is usually no effective prophylaxis or treatment available.

Again, the differences are about transmission to the community, with some mention of animal disease (although most HG-4 pathogens are essentially human diseases; although it could be argued, as their "normal" hosts are unknown, that these are in fact a collection of zoonoses). Again, this group is sometimes referred to as a *group of high individual* as well as *high community risk organisms*. All HG-4 organisms are viruses, without exception, although once again, it can be argued that some of these organisms pose lesser risks to laboratory workers than some HG-3 microorganisms. This is because some of these viruses are of low infectivity via the respiratory route (e.g., Crimean Congo hemorrhagic fever virus; CCHFV) and many are blood-borne pathogens. This is, of course, not true for all of these viruses; some are highly infectious from aerosols (e.g., Lassa virus) and pose a great risk to laboratory and healthcare workers (Figure 2.1). Here, as with the other HGs, we see that a thorough understanding of the biology of the microorganism, in particular, its route of transmission is essential before undertaking a risk-based categorization of these pathogens.

Most countries have a panel of experts that meet regularly to assign HGs to emerging pathogens. In the United Kingdom, the Advisory Committee on Dangerous Pathogens (ACDP) makes these recommendations. Because the procedures for assigning an HG to a newly emerged pathogen vary between countries, it is highly likely that countries will assign the same pathogen to different risk groups. For example, a comparison of the risk groups assigned to bacterial pathogens by Western governments can be seen at *http://www.absa.org/riskgroups/bacteria.html*.

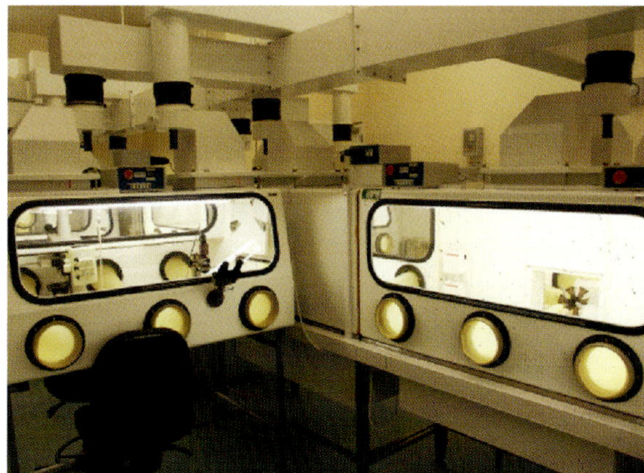

Figure 2.1 A cabinet line laboratory for work with hazard group 4 pathogens.

2.1 Facility Requirements

The containment facility requirements for handling pathogens within the different HGs largely follow the pathogen categorization, for example, HG-3 organisms would require a biosafety level 3 (BSL-3) laboratory (Figure 2.2).

However, this general rule does not always hold true as the HG only gives an initial guide to which BSL of laboratory is required to work with a particular infectious disease agent, but a full risk assessment needs to be performed as well. The Biosafety in Microbiological and Biomedical Laboratories (BMBL fifth edition, 2007) gives specific advice on the containment levels required for handling both diagnostic and research materials. The BMBL recommends that **diagnostic** samples of many of the HG-3 agents may be handled in BSL-2 laboratories, but growing samples in large quantities required BSL-3 facilities (e.g., *Bacillus anthracis*). For other agents such as *Brucella* spp., clinical tests may be performed at BSL-2 but cultures should always be performed at BSL-3. Table 2.1 summarizes the recommended safety equipment required to handle microorganisms within the four BSLs (BMBL fifth edition, 2007).

An example of the risk assessment would be that the work proposed results in the generation of high-concentration aerosols, then BSL-3 may be required to provide the necessary degree of operator protection, as it ensures superior containment of aerosols in the laboratory workplace. The BSL selected for the specific work to be done is decided on by professional judgment based on a comprehensive risk assessment, rather than by automatic assignment of a laboratory Biosafety Level according to the particular risk group designation of the pathogenic agent to be used (Figure 2.3).

In the United Kingdom, the interpretation is currently different, although new regulations currently awaiting government ratification will change this; risk group 3 agents are currently handled in containment level 3 laboratories, unless there is

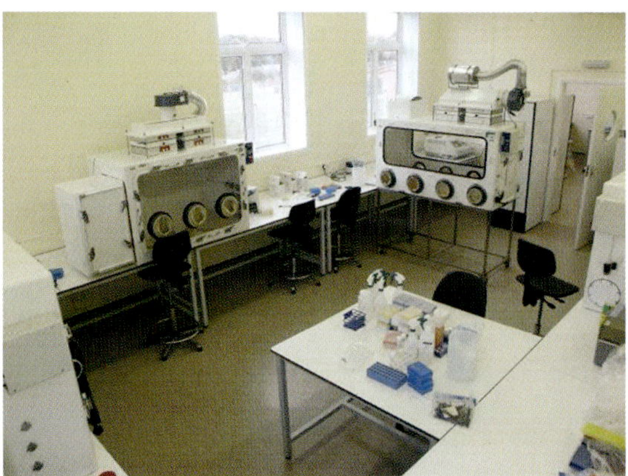

Figure 2.2 A UK-based BSL-3 laboratory, using class III biosafety cabinets.

Table 2.1 Summary of recommended biosafety levels for infectious agents.

Biosafety level	Agents	Practices	Safety equipment (primary barriers)	Facilities (secondary barriers)
1	Not known to consistently cause disease in healthy adults	Standard microbiological practices	None required	Open bench top, sink required
2	Associated with human disease, hazard = percutaneous injury, ingestion, mucous membrane exposure	BSL-1 practices plus: limited access biohazard warning signs "Sharps" precautions Biosafety manual defining any needed waste decontamination or medical surveillance	Primary barrier = class I or II BSCs or other physical containment devices used for all manipulations of agents that cause splashes or aerosols of infectious materials; PPEs; laboratory coats; gloves; and face protection as needed	BSL-1 plus: autoclave available
3	Indigenous or exotic agents with potential for aerosol transmission; disease may have serious or lethal consequences	BSL-2 practices plus: controlled access decontamination of all waste decontamination of laboratory clothing before laundering baseline serum	Primary barrier = class I or II BSCs or other physical containment devices used for all open manipulations of agents; PPEs; protective laboratory clothing; gloves; and respiratory protection as needed	BSL-2 plus: physical separation from access corridors self-closing, double-door access exhausted air not recirculated negative airflow into laboratory
4	Dangerous/exotic agents that pose high risk of life-threatening disease, aerosol-transmitted laboratory infections, or related agents with unknown risk of transmission	BSL-3 practices plus: clothing change before entering shower on exit all material decontaminated on exit from facility	Primary barriers = all procedures conducted in class III BSCs or class I or II BSCs *in combination with* full-body, air-supplied, positive pressure personnel suit	BSL-3 plus: separate building or isolated zone dedicated supply and exhaust, vacuum, and decontamination systems other requirements as outlined in BMBL fifth edition

Abbreviations: BSC, biosafety cabinet; PPE, personal protective equipment.
(Reproduced from BMBL Fifth edition, 2007.)

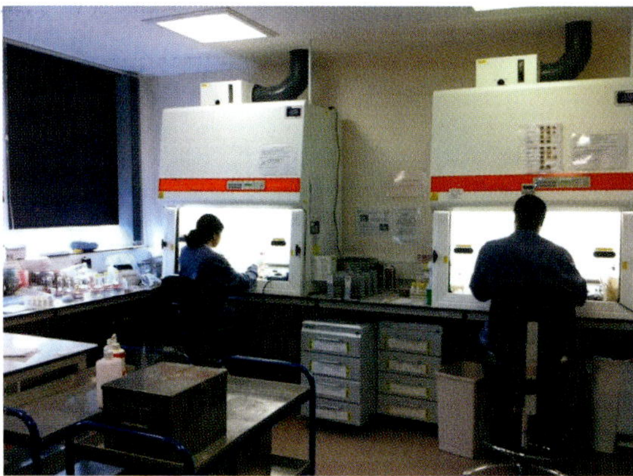

Figure 2.3 A typical BSL-3 diagnostic laboratory, using class II biosafety cabinets.

a derogation for low hazard clinical work that applies to agents that are infectious by the oral or blood-borne route.

2.2 Exceptions to the Rules

There are certain circumstances where there are exceptions to these rules and where local risk assessment may not be completely adequate. For example, some HG-2 agents are termed as *2+* and represent an increased risk to the laboratory worker, usually via the aerosol route. These microorganisms are always handled using an appropriate microbiological safety cabinet and examples of these pathogens are *Legionella pneumophila* and *Neisseria meningitidis*. Nontoxigenic variants of some pathogens may be effectively "downgraded" by application to the regulatory authority, likewise vaccine strains usually fall into this category, for example, *B. anthracis* Sterne strain.

Categorization does, of course, not consider the work that is actually being carried out using a pathogenic microorganism. Such factors include the volume being handled, the titer of the material, the procedures undertaken, and the zoonotic characteristics of the microorganism, what the normal route of infection is for a particular pathogen (e.g., if the pathogen is a vector-borne disease, then vector control is important but the direct risk to the laboratory worker may be relatively low). Other considerations for determining an appropriate facility BSL include the security measures required (Select Agent Rule; Schedule 5 list, Australia Group; etc.), whether animals are involved in the studies and whether any of the microorganisms are genetically modified organisms. The risk assessment needs to consider all of these factors before a Biosafety Level for handling a pathogen is finally assigned. The risk assessment should also be a dynamic document and modified if new evidence emerges to suggest that additional factors need to be considered.

Human factors that also need to be considered are as follows:

- preexisting disease
- compromised immunity (e.g., important when working with *Vaccinia* virus)
- pregnancy
- effects of medication (a consideration when *Clostridium difficile* is being handled in a laboratory).

These human factors may provide additional risks to laboratory workers, depending on the work being undertaken and need to be considered as a part of the risk assessment.

Summary

There are four defined HGs used worldwide, which categorize pathogenic microorganisms based on the following criteria:

- whether the agent is pathogenic for humans and/or animals;
- whether the agent is a hazard to employees;
- whether the agent is transmissible to the community;
- whether there is effective treatment or prophylaxis available.

3
Technical and Practical Aspects of BSL-3 Laboratories

Frank T. Hufert and Manfred Weidmann

The primary risk criteria, which define the biosafety levels (BSLs) 1–4, are infectivity, severity of disease, transmissibility, and the nature of the work being performed and, especially in veterinary situations, whether the microorganism is indigenous or exotic.

The majority of pathogens used in human medicine and in veterinary laboratories can be handled under BSL-1 and BSL-2 conditions. However, some pathogens need handling using the higher BSL-3 or BSL-4 laboratory containment level to protect the environment and the laboratory workers.

The classification of the different microorganisms and viruses is in part defined within the European Union [1] and on the level of the different EU member states (e.g., Germany [2]).

In this chapter, we focus on some technical and practical aspects of working in BSL-3 facilities.

Local regulations may vary in different countries. Thus, for building a BSL-3 or BSL-4 facility, a very strong collaboration between the local authorities, the engineering and planning companies, the fire brigade, and the user is necessary. The very early interaction of this network, especially the communication with the user, is the key feature and a very important prerequisite to achieve the construction of a cost effective, safe, and usable BSL-3 facility.

3.1
Technical Aspects – Facilities, Secondary Barriers

The BSL-3 facility is mainly built to protect the environment from an organism or a genetically modified microorganism classified into hazard group 3. Thus, the laboratory must be constructed to avoid the escape of microorganisms. This can be achieved by integrating the facility into an existing building or by setting up a new building. It is of importance that the facility is placed on the first floor or further up because this protects it from flooding that might occur due to heavy rainfall. In principle, it is best to have the virology laboratory completely separated from the microbiology laboratory to avoid any cross contamination. In case that only

Working in Biosafety Level 3 and 4 Laboratories: A Practical Introduction, First Edition.
Edited by Manfred Weidmann, Nigel Silman, Patrick Butaye, and Mandy Elschner.

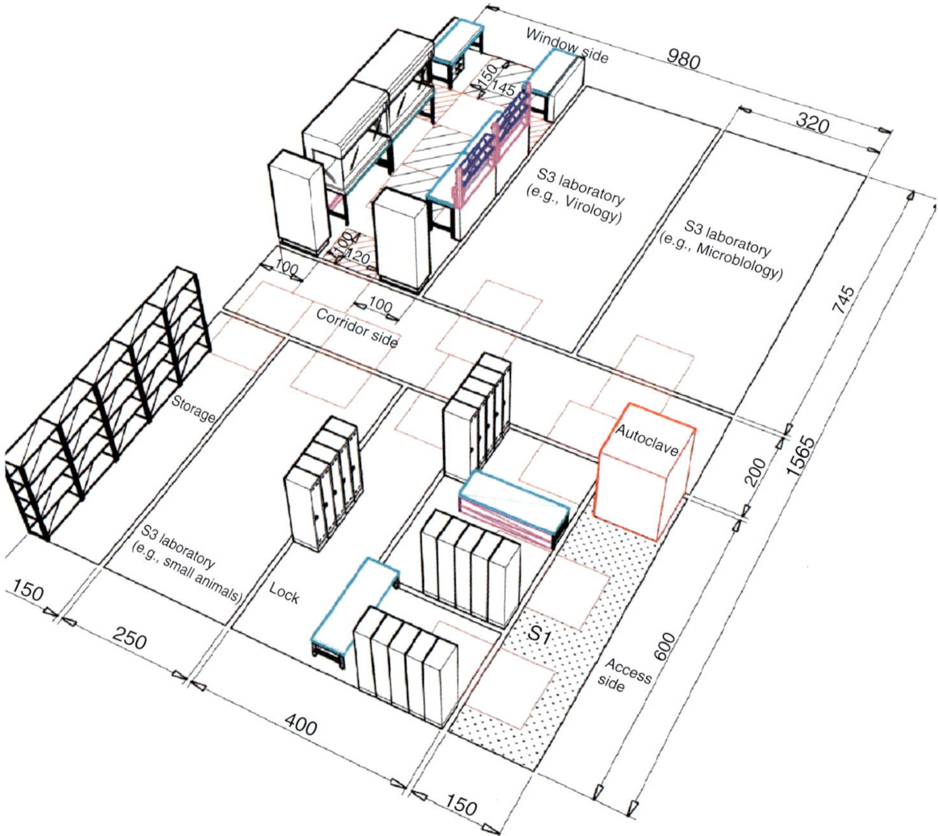

Figure 3.1 Principle construction of a BSL-3 facility. Access is through two separate locks for female and male workers (bottom right). They reach through autoclave opens into the anteroom. An example of four laboratories is given, one of which could be a room for small animal experiments or for large laboratory equipment such as freezers and centrifuges. A storage room for laboratory plastic ware, gloves, autoclave bags, respirator filters, and so on is often not planned but a good idea. (The construction plan was kindly provided by Volker Krieger, Consulting Services, Freiburg, Germany.)

one facility is available, a strict separation of virology and microbiology laboratory rooms is mandatory.

The basic principles of BSL-3 laboratory construction are illustrated in Figure 3.1. Compared to a BSL-2 laboratory, the BSL-3 laboratory has additional requirements that include:

1) physical separation from access corridors,
2) self-closing door access,
3) exhaust air system without recirculation,
4) negative airflow into laboratory.

The laboratory needs a single access lock for entry of staff, corridors, storage rooms, and laboratory rooms where the work with infectious agents is carried out. The protection of the environment in these facilities is achieved by permanently closed windows and air filtration systems generating a negative pressure gradient. The gradient of the negative pressure decreases from the lock to the laboratory rooms, generating a continuous influx of air into the facility and the hot zone of the laboratory where the work is carried out. All doors to the different rooms should close automatically and must be opened only to enter the individual rooms.

Furthermore, any type of waste has to leave the laboratory via an autoclave, best constructed as a double-door-autoclave. Any wastewater also needs to autoclaved or otherwise treated for disinfection before leaving the laboratory. All technical systems have to be controlled electronically by sensors and technical faults must be recorded and reported immediately to well-trained technical staff.

3.1.1 Air Filtration Systems

Incoming air and outgoing air should be filtered using modern air filter systems that use HEPA filters (high-efficiency particulate air) to remove particles. They are typically made from paper-thin borosilicate medium, pleated to increase surface area, and fixed to a frame. Different classes of filters are used. Incoming air should at least be filtered using H7 filters according to DIN/EN1822, whereas outgoing air must pass at least an H13 filter that reaches an efficiency of 99.95% with an overall penetration value of 0.05%. However, it is not necessary to filter incoming air at that level. Prefiltration with an H7–H10 filter extends the lifetime of the HEPA filters on the exhaust side and additionally it keeps the laboratory clean (no dirt particles).

All filter types are constructed as bag-in/bag-out systems, and the bagged filters whether the filter units are placed in the laboratory (Germany) or outside (the United Kingdom) must fit into the autoclave for disposal. Furthermore, the different ventilation lines should have individual ventilation flaps to allow change of filters easily without having to shut down the whole laboratory when changing an individual filter.

Filter integrity must be checked on an annually basis to detect potential leakage and should be performed using semiautomated scanning technology [3]. Before changing filters, they must be decontaminated, optimally by fumigation with formaldehyde gas. All decontamination procedures must be validated using biological and chemical indicators and after filter replacement scan testing of filter integrity must be performed (Figure 3.2).

3.1.2 Water

The number of taps and sinks should be reduced to the minimum needed, as they are error prone and might cause accidental flooding of the laboratory. Thus, sensors to detect water on the floors must be installed at the appropriate place

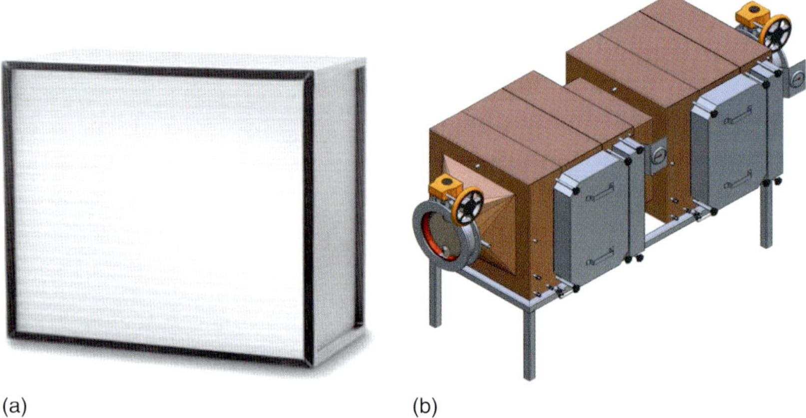

(a) (b)

Figure 3.2 Filter. (a) H13 filter and (b) double filter system with access points for automated or semiautomated integrity scanning option. (*http://www.camfilfarr.com/cou_uk/*; Foto: Camfil.)

and height. All sinks must drain their effluent into containers or into an effluent treatment plant, which then has to be autoclaved before draining it into the public sewage system. A double-door autoclave is best for handling individual sewage containers. To avoid overflow of the sewage containers, they also should be sensor controlled.

The amount of wastewater to be autoclaved increases dramatically when air-conditioning systems are installed inside the laboratory. Thus, it is more convenient and practical to chill down air outside the laboratory before it passes the first filter of the laboratory ventilation system. The energy input for precooling may even be lower than the energy input for air conditioners and additional autoclaving of condensate.

3.1.3
Fire Protection

A BSL-3 laboratory must be constructed using the F 90 standard (fire resistant for 90 min, burn down strategy) for all windows, walls, floors, and ceilings. Smoke sensors must be installed in all rooms. The system used for fire control is always a matter of debate and must be discussed according to the local regulations with representatives of the fire brigade and the corresponding authorities. In principle, every room must be equipped with a fire extinguisher for rapid use to immediately allow firefighting when laboratory staff are present. However, there is a need for a fire control system to monitor the laboratory when users are absent. We do not recommend an automated extinguishing system using gas because there is the danger of suffocation in the case of false alert. Wet (i.e., water-filled) sprinkler systems might also be error prone and lead to accidental flooding of the laboratory because of a technical fault or a false alert. Thus, we recommend a semidry sprinkler system that is filled by the fire brigade on arrival with sprinklers opening only if the room temperature exceeds 70 °C, an indication for a large fire. This is the system with the lowest error rate only acting on demand by the fire brigade. All technical containment measures try to reduce the amount of aerosols in the laboratory therefore aerosol-producing nebulizing sprinklers are only the second best choice, especially in the case of a false alarm.

In addition, as biological agents are rapidly heat inactivated, there is no danger in using this type of system, and it provides a chance for a sortie into the laboratory by a fire brigade team in the case of a small fire.

All individual floors must be equipped with water barriers to avoid any flooding of adjoining parts of the building in case the sprinkler system or water-based extinguishers are used.

3.2 Practical Aspects – Safety Equipment, Primary Barriers

3.2.1 Staff

Only experienced staff should work in a BSL-3 facility and people should be trained for working in a BSL-3 laboratory. Training courses however are few but should be mandatory for everybody working in a BSL-3 or BSL-4 laboratory. Some additional training courses can be found here [4–6]. The installation of a video system for safety and monitoring of staff is a good option to improve biosecurity. Electronic access is also recommended.

3.3 Personal Protective Equipment (PPE)

Laboratory workers in a BSL-3 facility must wear protective laboratory clothing, goggles, gloves and/or gauntlets, shoes, and respiratory protection as needed. After entering the lock, they have to change clothes and wear laboratory shirts and trousers as well as two laboratory coats before entering the BSL-3 facility. Three pairs of latex gloves protect the laboratory worker. The first pair of short gloves is taped onto the sleeves of the first laboratory coat and the second pair of gloves should be gauntlets pulled over the second laboratory coat covering the arm up to the elbow. This protects the cotton gown arm sleeves from the downflow air of the biosafety cabinet (BSC) class II potentially carrying infectious aerosols. Gauntlets can be easily disinfected after finishing work in the BSC or discarded if need be.

The third pair of short gloves is used in the hot zone of the BSC. It is discarded in the BSL-2 cabinet immediately after working and disposed in the waste container inside the BSL-2 cabinet. This routine avoids any accidental export of hot infectious material on a glove.

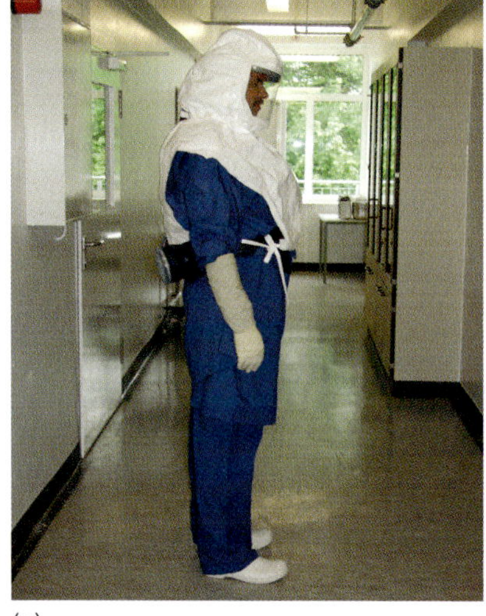

(a)

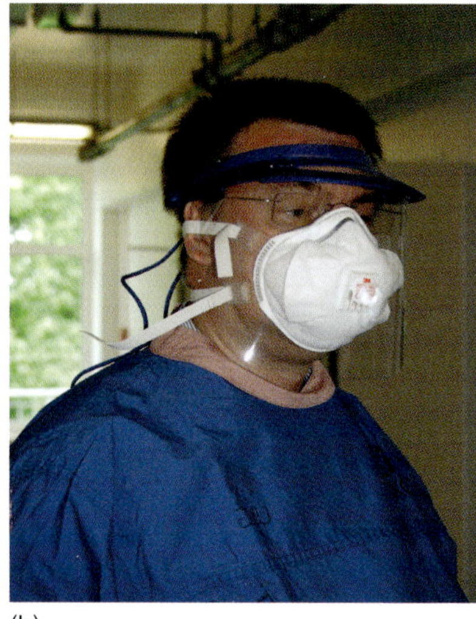

(b)

Figure 3.3 Personal protective equipment including respiratory protection. Personal protection equipment including protective laboratory clothing, long gloves covering the arm up to the elbow, and respiratory protection. (a) Example of respiratory protection using a HEPA-filtered positive pressure mask. (b) Respiratory protection using an N95/FFP3 mask combined with a face shield.

Respiratory protection is needed in case that aerosol transmission of the agent is known. Two types of protection are common: (i) positive HEPA-filtered pressure system with head and shoulder hood (Figure 3.3a) or a smaller lighter hood (not shown) and (ii) N95/FFP3 mask in combination with a face shield as shown in Figure 3.3b.

3.3.1 Primary Barriers and Working Procedures

Any work with infectious material in the BSL-3 facility must be performed in a BSC class II or III type to protect the user from events that might cause splashing, spraying, or splattering of droplets or the generation of aerosols. BSCs should also be used for the initial processing of clinical specimens, when the nature of an agent is likely to be transmissible by aerosol (Chapter 10). The principles of BSCs type II and type III are illustrated in Figure 3.4. Class II cabinets can be divided into types A and B on the basis of construction type, airflow velocities and patterns, and exhaust systems [7, 8]. They are made for product, personnel, and environmental protection to handle microorganisms in BSL 2–4 facilities.

In contrast to class II cabinets, class III cabinets are totally closed systems, gas tight with HEPA-filtered supply, and exhaust air. They are designed as glove boxes where work is performed with attached long-sleeved gloves. The cabinet is kept under negative pressure of at least 120 Pa and airflow is maintained by an

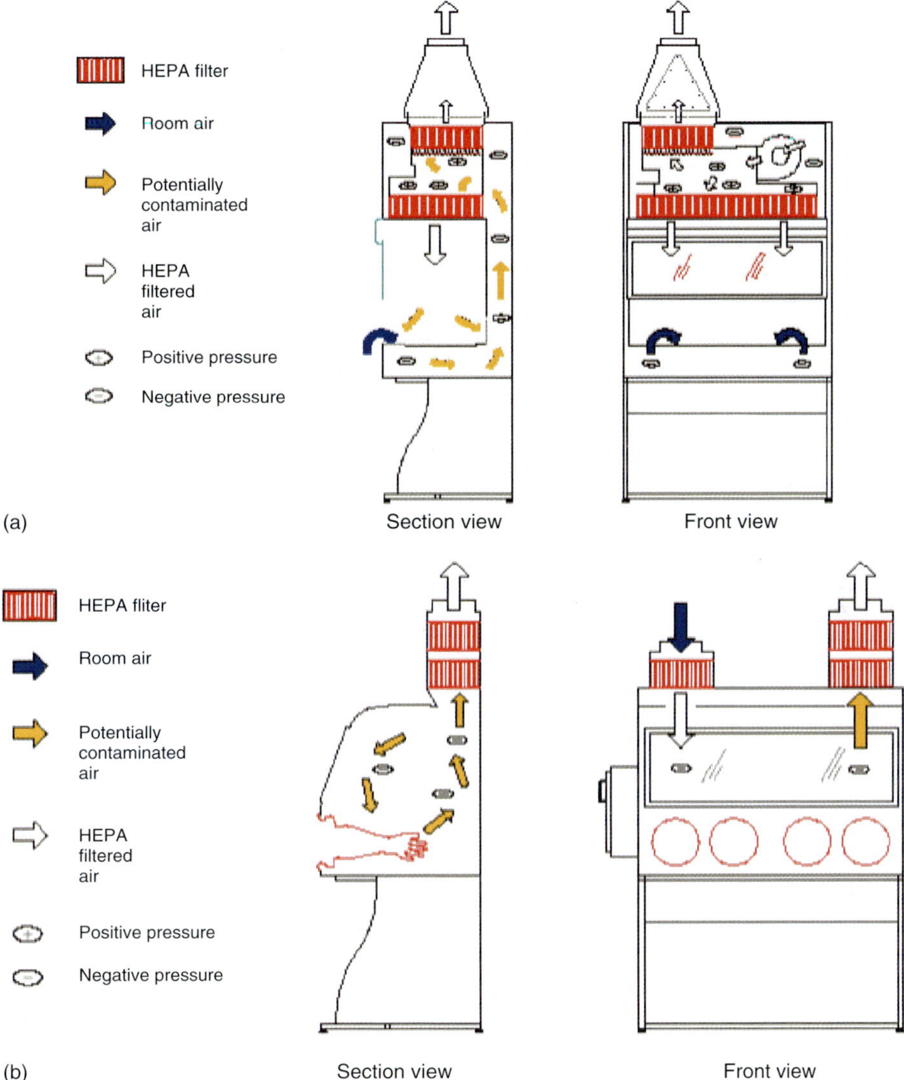

Figure 3.4 (a) Biosafety cabinet class II type A1. (b) Biosafety cabinet class III type (glove box). (From [8], Reproduced with permission from the Minister of Health, 2013, *www.publichealth.gc.ca*.)

exhaust system. Class III cabinets are made for protecting the worker and the product. The exhaust air is double HEPA filtered. Any removal of materials from the cabinet must be through a dunk tank, double-door autoclave, or air lock pass through for decontamination. Technical prevention of simultaneously opening both lock sides is mandatory. BSC class III cabinets are designed for working with level 4 pathogens and provide an alternative to the positive-pressure suit made for maximum containment laboratories. Especially for diagnostic laboratories, this system is an economical solution to handle samples suspected to contain a BSL-4 agent to prepare them for molecular diagnostics. Furthermore, the BSC class III cabinets are also recommended for work with highly pathogenic spore-producing bacteria, such as *Bacillus anthracis*, to increase the level of protection and to avoid accidental release of highly stable spores into the laboratory, especially in small units when no separate laboratory rooms are available to separate virological and microbiological works.

In the case of BSC class II work, we recommend using BSCs of 1.80 m width only. They provide a comfortable working space and simultaneously ample space for solid and fluid laboratory waste containers inside the bench. Liquid waste is collected separately from solid waste in containers containing an aldehyde-based (or other validated) disinfectant. To allow air circulation around the waste containers and to avoid airflow turbulences inside the BSC, they are elevated by special racks, as shown in Figure 3.5.

During the bench, work hands should stay in the cabinet at all times and the third pair of gloves is to be removed and disposed of in the waste container when bench work is finished, followed by disinfection of the gauntlet up to the elbows.

It is mandatory that all centrifugation steps must be carried out in closed system centrifuges only. It is not a good idea to run centrifuges inside a BSC, as the turbulence caused during centrifuge operation completely upsets the controlled laminar airflow. The same is true for Bunsen burners.

To remove tubes from the bench, the closed tubes should be disinfected externally using an aldehyde-based disinfectant and passed on to a second worker assisting the bench worker. All waste containers should be closed and removed from the BSC and autoclaved immediately after work has finished. Bench disinfection is mandatory after any type of work. During bench disinfection, special attention should be paid to the air inlets at the front of the bench because they are potentially highly contaminated with infectious material, as all the contaminated air has passed

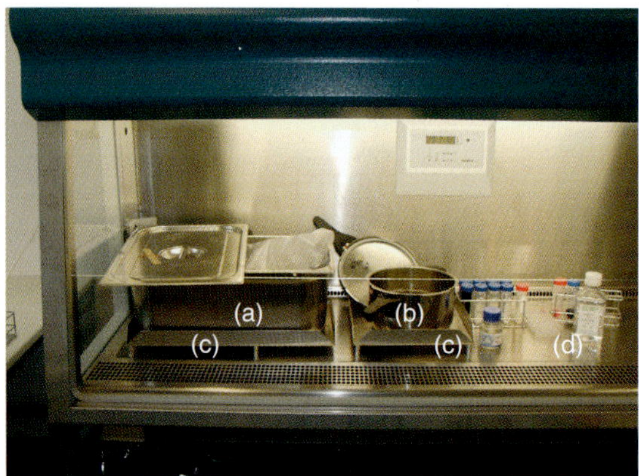

Figure 3.5 Working setup of a class II BSC. (a) Standard waste container, (b) liquid waste container, (c) rack to allow air circulation around the containers, and (d) working area.

through them throughout the working session. However, we do not recommend UV light for microbe inactivation inside the BSC because the microbe inactivation efficiency of UV light decreases rapidly (Chapter 7).

After working, all racks and other nondisposable gear are chemically disinfected for at least 1 h using a benchtop bath containing an aldehyde-based disinfectant.

Before leaving the laboratory, all workers have to change clothes and the laboratory dress is disposed to be autoclaved.

Summary

In this chapter, we have given you some practical advice to set up a BSL-3 facility. However, keep in mind that maintenance costs have to be paid after you have switched on the biocontainment facility. A maintenance budget, research grants, and permanently employed and experienced staff are a prerequisite to run a BSL-3 laboratory successfully.

References

1. (2000) Directive 2000/54/EC of the European Parliament and of the Council of 18 September 2000 on the protection of workers from risks related to exposure to biological agents at work. *Off. J.*, **L 262**, 0021–0045.
2. Verordnung über Sicherheit und Gesundheitsschutz bei Tätigkeiten mit biologischen Arbeitsstoffen (Biostoffverordnung – BioStoffV) vom 27. Januar 1999 (Bundesgesetzblatt Teil I, S. 50), die zuletzt durch Artikel 3 der Verordnung vom 18. Dezember 2008 (Bundesgesetzblatt Teil I, S. 2768) geändert worden ist.
3. Camfil *http://www.camfilfarr.com/cou_uk/filtertechnology/filtertesting/?m=3* (accessed 13 April 2013).
4. Health Protection Agency *http://www.hpa.org.uk/EventsProfessionalTraining/InfectionsTrainingAndEvents/InfectionsTrainingCourses/trainbiosafetyprinciplescontainmentlevel3/* (accessed 13 April 2013).
5. Institut Pasteur de Lille *http://www.pasteur-lille.fr/fr/formation/outils_scientifiques/index.html* (accessed 13 April 2013).
6. University Medical Center Götingen, Department of Virology *http://www.virologie.uni-goettingen.de/index.php?page=22&empty=1&id=26* (accessed 13 April 2013).
7. Public Health Agency of Canada *www.publichealth.gc.ca* (accessed 13 April 2013).
8. Public Health Agency of Canada (2004) Biological safety cabinets, in *The Laboratory Biosafety Guidelines*, 3rd edn Chapter 9.

4
Animal Biosafety Level 3 Facility – Enhancements When Dealing with Large Animals

Francesc Xavier Abad, David Solanes, and Mariano Domingo

Most livestock pathogens can be safely studied in biosafety level (BSL)-2 or -3 laboratories and *vivaria* [1] using small laboratory animals (mice, rats, and guinea pigs).

However, when dealing with certain highly pathogenic livestock agents, especially zoonotic pathogens, facilities have to be enhanced by setting up several modifications designed to protect the environment. These include placing animals in isolation in primary containment containers with high-efficiency particulate air (HEPA) filtration of supply and exhaust air, and effluent decontamination. The secondary barrier, then, consists of controlled staff entry and exit through a dressing and shower system, double-door autoclave, and/or air lock, decontamination systems (or a fumigation chamber) (Figure 4.1).

As all work with infectious materials is performed within primary containment equipment (primary barrier), a facility integrity test (e.g., pressure decay test) is not required.

Large animals (pigs, cattle, sheep, goats, horses, and also wildlife animals such as deer, chamois, and Spanish ibex) cannot be placed inside any kind of animal isolation device (Figure 4.2). In this case, the structure and the procedures of the facility and the engineering systems, which generally act as a secondary barrier, have to be transformed into a primary barrier (or containment).

The facilities then have to fulfill large animal biosafety level 3 (LABSL-3), ABSL-3-plus (animal biosafety level), or ABSL-3Ag (*Ag* brief for *Agriculture*) requirements [2–4]. It has to be clearly stated that as in all other BSL-3 facilities, the engineering and the procedures of the LABSL-3 facilities slightly enhance worker safety but mainly are used for protecting the environment from high-risk pathogens.

The escape and spread of such high-consequence pathogens could have a major impact on the animal health status of a country, in terms of morbidity and mortality of livestock, and further international trade implications. The outbreaks of foot-and-mouth disease in the United States in 1978 and in the United Kingdom in 2007, serve as an example of the aforementioned.

It is out of our scope to review all the enhancements that can be done in ABSL-3 facilities dealing with small animals (mice, rats, guinea pigs, etc.). In this

Working in Biosafety Level 3 and 4 Laboratories: A Practical Introduction, First Edition.
Edited by Manfred Weidmann, Nigel Silman, Patrick Butaye, and Mandy Elschner.

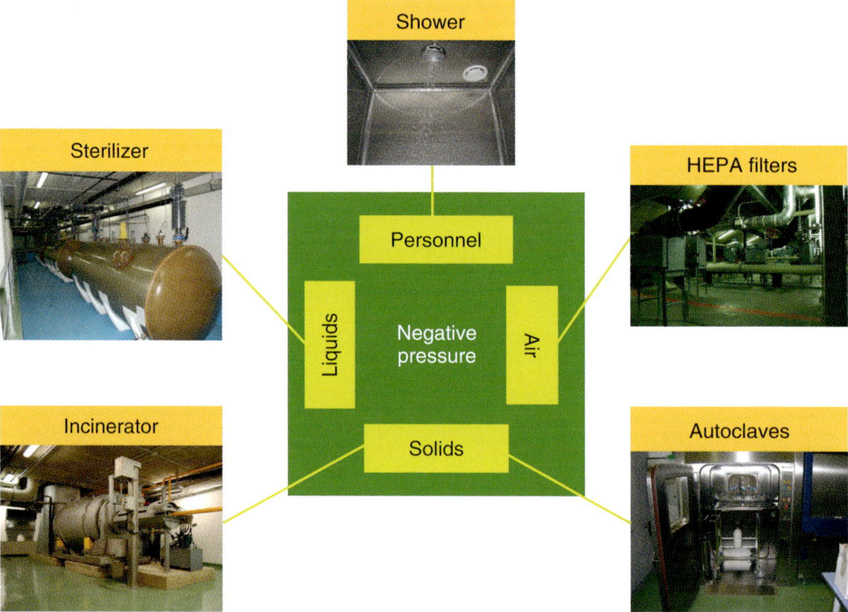

Figure 4.1 ABSL-3 facility. (Provided with the permission of CReSA.)

case, several kinds of primary containment devices, from static filter-top cages to systems designed for individual ventilated housing providing a sealed containment environment at the cage level, are available. Handling of animals in these cages has to be performed within biological safety cabinets to minimize the potential aerosol release.

In this chapter, we describe an ABSL-3 setup for handling large animals (LABSL-3). In this situation, facility barriers, which normally act as secondary barriers, must be considered as primary barriers. For small animals, the standard BSL-3 containment features are enough. When the animal facility has to manage large animals, these features are only the starting point. Several characteristics ordinarily used in a BSL-4 facility must be included as enhancements.

For LABSL-3 facilities, all animal rooms (or boxes) must be designed, constructed, and certified as primary containment barriers. All these boxes placed inside a larger area also act as a BSL-3 unit (which can include several laboratory areas) in order to follow the principle of *a box in a box*. As the needs of HEPA-filtered air and solid and liquid waste treatments can be very high, the box in a box system has to be placed into a sandwich structure, with a lower level intended for waste treatments and an upper floor dedicated to air filtration systems (Figure 4.3) [5].

In LABSL-3 facilities, only the zoonotic potential or the pathogen may modify the rules in place. For non-zoonotic pathogens, the containment is identical for animals and workers. For the zoonotic ones, the staff must be protected by using specific personal protective equipment (PPE) such as respiratory protection.

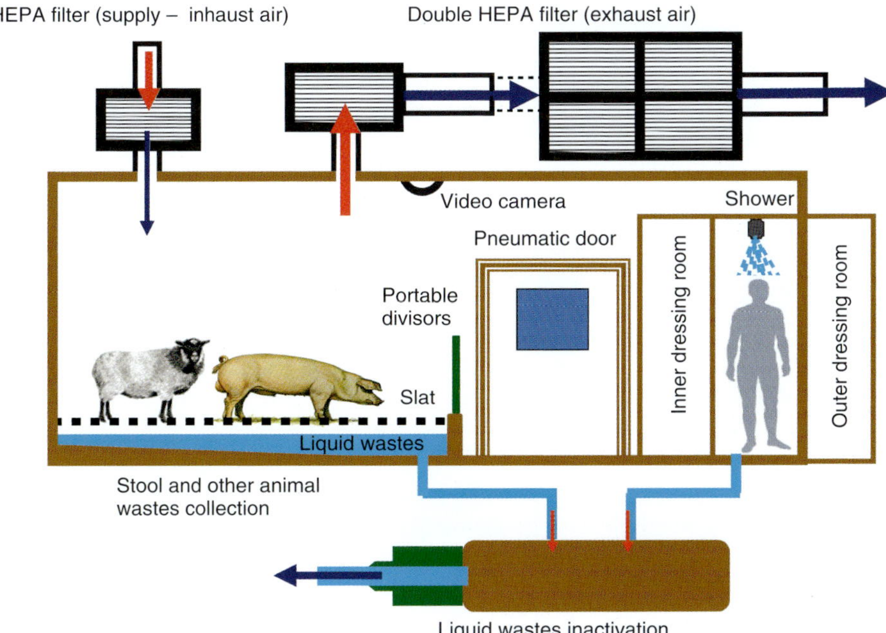

Figure 4.2 A schema of an animal box. Floor drains have to be designed with a sufficient slope to avoid the pooling of water.(Provided with the permission of CReSA.)

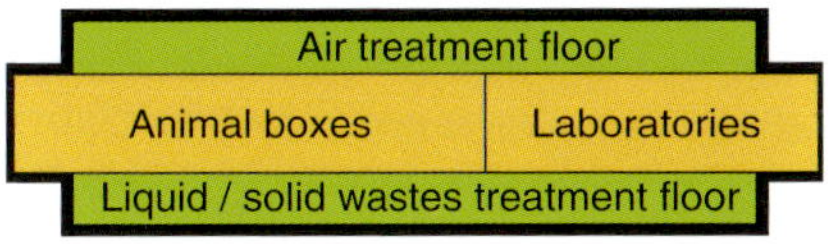

Air treatment floor

Animal boxes | Laboratories

Liquid / solid wastes treatment floor

Figure 4.3 A box within a box in a sandwich structure. The concept "box within a box" suggests that the containment area or hot zone is always surrounded by a secondary wall or protection from the outside. However, it is a flexible concept and each facility has to find its way.

4.1 Enhancements to Upgrade a Standard Animal BSL-3 Facility to a LABSL-3 Facility Housing Large Animals

1) The entry and exit of personnel in the whole LABSL-3 facility is restricted, by electronic access. All personal clothing, including underwear and personal belongings, must be removed in the outer dressing room. All persons entering the laboratory must pass through a shower system or airlock (both with interlocking doors) and then access to the inner dressing room where they must wear laboratory clothing.
2) The entry and exit of the personnel into an animal box (located inside the perimeter of the whole facility) can only be done through a series of rooms (Figure 4.2):

a. an outer dressing room, where personnel have to remove all their clothes, including underwear;
b. a shower room, ventilated with compressible gaskets on both doors, preferably interlocked; procedural shower can be mandatory at the exit, at the entry and the exit, or only at the entry. The shower practice in force has to be strictly followed and it consists of a full body shower using soap;
c. an inner (considered as dirty) dressing room, within the animal box with specific work clothing that has to be worn by the personnel before entering the animal holding area. These clothes can be used for several days for animal care personnel but at the end must to be collected, closed in hermetic containers, and autoclaved before being laundered, at least when dealing with zoonotic agents.

3) Not only material, equipment, and supplies but also the animals have to enter the ABSL-3 only through air locks, SAS (sterilized air system), fumigation chambers, or similar devices [3]. All these systems have to be provided with air-inflated or compressible gaskets and interlocked doors. The compressed air lines serving the air-inflated gaskets must be provided with HEPA filters and check valves. Alternatively, an interlocked double-door autoclave can be used for small materials. Biological materials should not enter through showers.
4) The transfer of material within the facility must be done in several ways:
   a. from clean areas as laboratories or SAS air lock to the animal boxes, a pass-through liquid dunk tank can be used; alternatively, a shower system is acceptable;
   b. from animal boxes to BSL-3 laboratories through a dunk tank (for instance, full of chloramine-T, contact time of 30 min) or any other airtight-sealed pass box;
   c. from animal boxes to necropsy room through an airtight door, in closed containers.
5) Directional airflows and negative air pressures within the LABSL-3 facility have to be provided and maintained at all times. This directional airflow moves from areas of less potential hazard toward the dirtiest or hot areas. Air supply and air exhaust systems have to be independent. Air supply should be used as conditioned air for containment spaces. In the event of power failure, an emergency system has to be in place to maintain the negative pressure – even at a lower intensity. An emergency generator should be in place, and the combustible for it should be stored on site making the facility independent of any local supply. An uninterruptible power supply (UPS) unit providing constant electric power during the transfer of power from normal to emergency power is also necessary. It must be capable of maintaining a constant voltage to ensure a constant operation and avoiding memory loss of data (for instance, set point conditions for environmental or biocontainment performance).
6) All incoming and exhaust air to and from the LABSL-3 facility has to be, at least, HEPA filtered. If the HEPA filters are in a technical floor inside the

facility (Figure 4.2) remember that several times during the lifetime of the facility, you will need to replace the silting filters and the autoclave size has to be compatible with these. If the HEPA filters are outside the biocontainment, they have to be located as close as possible to the containment space to reduce the length of potentially contaminated air ducts. In all cases, periodical checks to prove efficacy and containment ability have to be executed.

7) Double-door autoclaves have to be designed and installed to decontaminate laboratory waste. These double doors have to be interlocked in a way that the outer door can only be opened when the cycle has been successfully completed [3]. As the autoclave units are part of the biocontainment wall, they must have an airtight seal to the barrier wall. If it is possible, the body of the autoclave units can be placed outside the biocontainment envelope so that maintenance can be conveniently performed. Autoclaves located between *dirty* and *clean* areas within the LABSL-3 unit are also useful.
8) All inner surfaces of a LABSL-3 facility (walls, ceilings, and floors) and wall penetrations (pipes, ducts, etc.) have to be completely sealed to allow fumigation. Individual animal boxes need to be airtight for fumigation without affecting adjacent areas. External and internal windows have to be breakage resistant (indeed, bulletproof quality) and sealed.
9) Necropsy rooms have to be sized and equipped to process large animals (or very large, as the case of cows of half a ton, for instance). In addition, this sizing has to be extended to corridors, mechanical systems to move dead or sedated animals, animal box doors, and also materials and ways (practicable floors, fences, constrain systems, etc.) to keep the animals accommodated.
10) Incinerators must be designed and installed for the safe disposal of large carcasses and also solid animal wastes (contaminated bedding and residual food) of infected or noninfected animals. The incinerators have to be designed to work in a biocontainment unit as they are systems that offer a way to breach biocontainment through the chimney. For this reason, a forced negative pressure ventilates the chimney until it reaches a temperature of 300 °C. Modern incinerators have two chambers: a primary chamber with an ideal temperature of at least 800 °C and a secondary chamber (for smoke gasses) with a temperature of up to 1000 °C. Loads with high moisture content may lower the processing temperature. Emission of particulate matter and selected chemical contaminants must be considered. Alkaline hydrolysis (by using NaOH or KOH) in combination with high temperature (above 100 °C) in closed pressurized vessels is an interesting option. The final product of this tissue digestion is a solution that may be mixed with the effluents without problems and can be used as fertilizer, compost additive, ect.
11) All liquid effluents from the facility (from animal boxes, showers, toilets, laboratory sinks, floor cleaning procedures, and water from autoclave chambers) have to be collected and decontaminated in a central sterilization system before their disposal into municipal sewers (Figure 4.4) [3]. Chemical decontamination can be performed by using suitable chemicals which are not affected by the presence of high levels of organic matter. Two molar sodium

Figure 4.4 ABSL-3 facility: liquid wastes. (provided with the permission of CReSA.)

hydroxide can be used to achieve a pH above 12 in the complete volume to be inactivated. Incubation overnight is followed by neutralization with hydrochloric acid before final disposal. Hypochlorites are recommended in facilities working with foot-and-mouth disease. For small volumes, thermal decontamination systems are installed in many facilities; sometimes there is a combination of thermal and chemical inactivation. When working with large animals, liquid effluents drained from floors of animal boxes can be readily inactivated by a combined temperature and contact time process, usually in a continuous flow system (100 °C at atmospheric pressure for 20 min).

4.2 Additional Recommendations

- Be flexible. Design the animal boxes as multifunctional spaces by using floors made of plastic or metallic pieces, metallic divisors or fences, and so on, allowing workers to divide groups in the same box as a function of disease transmissibility (by direct contact, non-air-borne, vector-borne, etc.). Consider that, in the future, you may have to adjust to new requirements (forced by new sponsors, facility–government agreements, etc.) and the animal box has to be capable of housing multiple species over the life of the building.
- If you are expecting to usually deal with farm animals such as pigs or sheep, your air-lock systems need to be sized accordingly. If experimental activities embrace the use of large animals such as cows, special animals such as chamois and deer, or high numbers of farm animals (i.e., more than 50 pigs above 100 kg), a sheepfold or fenced space outside the facility has to be designed and put in place in order to be able to keep animals outside waiting for entry. If your experimental procedure involves handling 10 or 12 calves, it is impossible to transfer all of them in one single SAS operation into the facility when the air lock has been mainly sized for swine, for instance. This flexibility has to be extended, then, to the animal path through the facility until it reaches the animal box. Therefore, locate

the air lock (or SAS) for animal entry as close as possible to the corridor/s where animal boxes are located and, also, design wide enough doors in all animal boxes to allow an easy entry of large animals (at least 120 cm width and 220–250 cm height) and do the same for the corridors (150–250 cm width).

- Redundancy. *Redundancy* is defined as having more than one system supporting an individual mechanical function, usually elements linked to biocontainment environmental protection, staff protection, and scientific outcome protection. It would be wrong to assume that all mechanical systems of a LABSL-3 facility need to have redundancy. The design has to reflect a well-balanced weighting between costs and benefits.
- Supply and exhaust systems should be designed with prefilters to extend the lifetime of supply and exhaust HEPA filters (Figure 4.2). A set of easily replaceable prefilters can be installed within the animal box that can be replaced in the middle of an experimental procedure and always at the end of each experiment. In addition, installation of a second HEPA filter in the exhaust line will only marginally improve the efficiency. It will however provide greater containment confidence if the primary exhaust HEPA filter fails due to a mechanical defect (or if scheduled maintenance has to be performed) (Figure 4.2).
- Use big windows in the animal box doors and in laboratory areas to allow the checking of activities in adjacent boxes or laboratories. It also improves illumination, workers attitude, and biosafety as visual contact is maintained. However, they have to be sealed tight and shatterproof.
- Remember that animals are almost unpredictable. Therefore, design and construct all inner animal box surfaces (ceilings, walls, and floors) with durable material, which has to be moistureproof, seamless, nontoxic, and highly resistant to cleaning agents, scrubbing, and impacts.
- Floors should be resistant to the action of urine, hot water, and cleaning agents used at the end of an experimental procedure. They should support animals, racks, and so on, without becoming gouged, cracked, or pitted. Check that all door locks, containment systems, fence junctions, are animal-proof and cannot be opened, operated, or surpassed by any unexpected animal behavior or activity.
- Install a high-quality digital video camera system in each animal box to allow continuous scientific observation of infected animals and also to enhance occupational safety monitoring (Figure 4.2).
- A well-sized cold chamber (with compressor redundancy and enough cooling power) has to be designed and installed to be able to store and accumulate enough solid waste (carcasses and contaminated beddings) in leakproof sealed containers before loading the incinerator. In our case, a normal incinerator cycle of 8 h accounts for nearly 150 min of heating or cooling the incinerator. In other facility designs, carcasses are directly disposed by a directly ducted tunnel to the incinerator or alkaline hydrolysis system.
- If your facility design contains several floors (typically a sandwich system), hermetically sealed doors for avoiding air contamination have to be installed.

- A heavy-duty lift for goods and materials should be installed, for example, for the transport of carcasses or heavy items, in case your facility is distributed over several floors (sandwich concept).
- Do not forget that you have to store a lot of materials if you are performing varied work (spare animal cages, palletized food, animal bedding, animal restraints, cleaning materials, feeders, forklifts, etc.). Therefore, design an appropriate storeroom with an easy access to the entries of the animal boxes, in clean and dirty areas and, if possible, also near the lift.
- The electronic transfer of data and information from biocontainment spaces to the outside of the perimeter is highly encouraged. Although raw data on paper can reach the outside by fax, using data transfer helps to reduce paper, notebooks, pens, and so on inside the facility. High-quality digital photo and video cameras in laboratories and in each animal box are also desirable.

If you can obtain the budget for all of these recommendations, you are really a lucky person. If budget constraints prevail, choose the most important features for your facility.

Above all else, consider that the hard work really starts when you switch the facility on. It is almost always overlooked that a huge investment into a high-quality animal biocontainment facility will fail if there are no resources to keep it at the initial high level. The running costs of high biocontainment facilities (and among them, those dealing with large animals) are extremely high (energy, maintenance services, waste treatment, specialized staff, etc.) and rapidly exceed the initial construction, commissioning, and validation costs. Therefore, definitively, long-term funding for long-term programs should be guaranteed before embarking on planning a high containment facility.

Summary

Working with livestock pathogens in ABSL-3 laboratories requires an additional set of technical support and maintenance systems not needed for BSL-3 laboratories. ABSL-3 laboratories need a flexible concept to allow keeping of animals of different sizes and needs in a BSL-3 containment. This requires additional technical infrastructure for moving in animals and feed and for decontaminating manure and other soils as well as carcasses. The maintenance of experimental procedures and the decontamination of stables after experiments have finished ask for additional considerations in the biosafety routine.

References

1. National Institutes of Health (2007) *Biosafety in Microbiological and Biomedical Laboratories (BMBL)*, 5th edn.
2. Heckert, R.A. and Kozlovac, J.P. (2006) Special considerations for agriculture pathogen biosafety, in *Biological Safety: Principles and Practices*, 4th edn (eds D.O. Fleming and D.L. Hunt), Amer Society for Microbiology, Washington, DC.

3. Heckert, R.A. and Kozlovac, J.P. (2007) Biosafey levels for animal agriculture pathogens. *Appl. Biosaf.*, **12**, 168–174.
4. Manuel, J. (2008) Oversight without obstruction: the challenge for high-containment labs. *Environ. Health Perspect.*, **116**, A487–A489.
5. International Veterinary Biosafety Working Group (2006) *Veterinary Containment Facilities: Design and Construction Handbook* (eds P. Mani and P. Langevin), International Veterinary Biosafety Working Group.

5
Personal Protective Equipment

Nigel Silman

5.1
Definitions

Personal protective equipment (PPE) may be defined as "All equipment designed to be worn or held by a person at work to protect against one or more risks and any addition or accessory designed to meet this objective."

5.2
Regulatory Background

The regulations covering safe use of PPE come from a European Directive that covers the "minimum health and safety requirements" for any task to be carried out, and these requirements are driven by a robust risk assessment of the process to be undertaken. The use of PPE should be considered as a "last resort" because they protect only the worker who is wearing that PPE; engineering control measures (e.g., microbiological safety cabinets) protect at source and therefore everyone who enters that area is protected equally. In addition, if the PPE is worn or used incorrectly or is poorly maintained, then the individual protection factor accorded to the worker will be reduced by an unknown amount. Before a decision is made to use PPE for protection during a procedure, the following aspects should be addressed during the risk assessment process:

- **E** Eliminate the hazard – is there an alternative way of doing what is required?
- **R** Reduce the hazard – can lower amounts of the hazardous compound or biological agent be used?
- **I** Isolate the hazard – what engineering controls can be applied?
- **C** Control exposure/contact – can the contact time with the hazard be reduced?
- **P** PPE.
- **D** Discipline.

An easy way to remember this sequence during the risk assessment process is to use the mnemonic ERICPD as indicated earlier. As can be seen, PPE should

Working in Biosafety Level 3 and 4 Laboratories: A Practical Introduction, First Edition.
Edited by Manfred Weidmann, Nigel Silman, Patrick Butaye, and Mandy Elschner.

be considered only when the other options for reduction or replacement within a procedure have been assessed and discarded.

5.3 Routes of Entry and Types of PPE

The routes of entry for infectious diseases into the body occur via four routes. These are via the skin, via the eyes, by inhalation, and by ingestion (Figure 5.1).

Taking each in turn, entry via the skin can occur by contact with exposed areas of the skin and many chemicals are able to penetrate intact skin, although microorganisms are not. The appropriate forms of PPE to protect the skin include gowns and laboratory coats as well as gloves and gauntlets. Gowns and laboratory coats should be capable of being properly fastened – gowns fasten at the back to protect the user from spills on the front of the body and can be removed by "rolling off" the gown to encapsulate whatever is spilt down the front. Laboratory coats should be double crossed-over at the front of the wearer so that any spills cannot pass through onto clothing worn underneath. Sleeve protectors may be indicated when working in open-fronted cabinets [biosafety cabinet (BSC) class I and class II], particularly, if there is a high risk of aerosol generation during the procedure being performed. These over sleeves may then be discarded before removing the arms from the cabinet. Arrangements must be made to decontaminate the laboratory coats and/or gowns before laundering [usually by autoclaving if used in a BSL-3 (biosafety level) laboratory]. Laundering of these garments must never be performed "at home." Laboratory coats and gowns when used at BSL-3 must **NEVER** be worn outside the laboratory envelope to ensure that there is no cross-contamination of nontoxic environments.

Gloves (Figure 5.2) should be suitable for the task being undertaken; an example is the use of nitrile gloves in circumstances where chemicals are being used that are capable of passing through latex gloves. Gauntlets may be used in circumstances where protection is needed over the sleeve as well, for example, placing materials

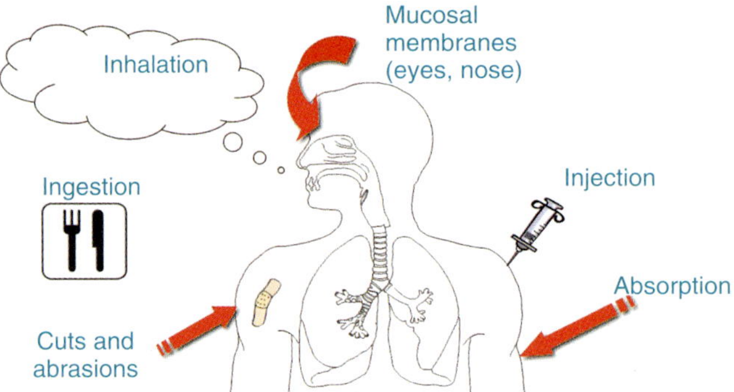

Figure 5.1 Routes of entry into the body that are used by infectious diseases.

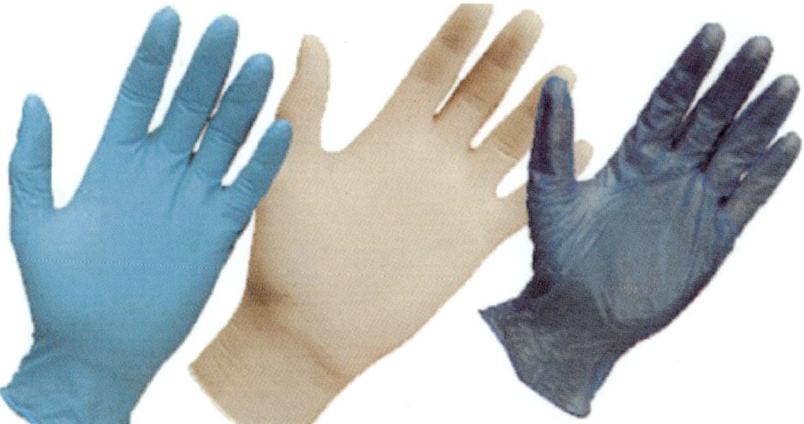

Figure 5.2 Different types of disposable glove used for laboratory work.

in chemical dunk tanks. The use of gauntlets here will prevent the chemical disinfectant from contacting the workers skin, whereas ordinary gloves will not. When working in open-fronted microbiological safety cabinets, two pairs of latex gloves should be used. The outer pair is considered "contaminated" and should be removed before removing the hands from the cabinet and discarded within the envelope of the cabinet. This way ensures that gloves that are worn within the laboratory do not contaminate other objects within the laboratory environment that may be touched with bare hands at any time.

Entry of infectious disease agents and chemicals may easily be prevented by the use of eye protection. In this category, consider using safety glasses, goggles, or a full-face visor. Safety glasses provide good protection against splashes that may be encountered during routine laboratory work; however, work with pathogens in a BSL-3 laboratory probably requires more specialized eye protection such as goggles that provide a seal around the eyes thus preventing ingress of chemical or biological agents. Full-face visors are useful in circumstances such as unloading autoclaves where hot liquids may be encountered or for viewing agarose gels under UV light (visors must be made from an appropriate UV protective plastic for this application). In all applications and during the risk assessment process, consider the suitability of these options for the particular hazard.

Infection via the inhalation and ingestion routes may be prevented by using engineering controls such as safety cabinets, fume cupboards, and powder stations as well as PPE such as positive pressure respirators and hoods (Figure 5.3). If engineering control solutions cannot be employed, then the use of respiratory protective equipment (RPE) should be considered (Figure 5.4). If these types of RPE are used, then the fit of the equipment should be tested (for respirators) as it is very easy to have leaks around the face seal with incorrectly fitted masks – this reduces the protection factor of such equipment considerably. The correct filter type for the hazard must also be used, as well as a means of determining how long the filter has been in use and how frequently they should be replaced. In most

Figure 5.3 Positive pressure hood.

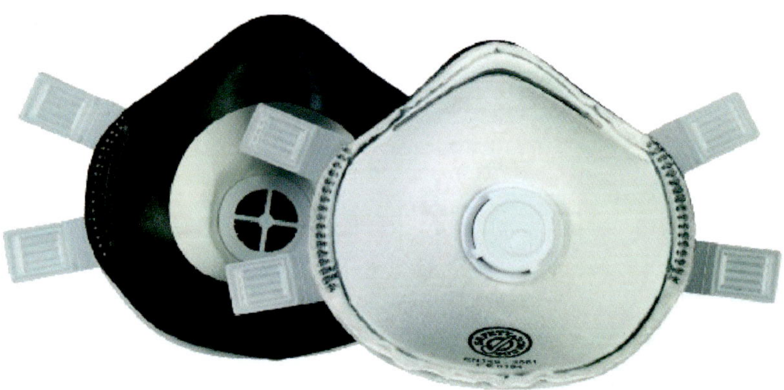

Figure 5.4 FFP3 (filtering face piece) dust mask.

instances, filters are used that provide both protection against particulate biological agents and having chemical scrubbers, so that they may be used, for example, post-formaldehyde fumigation of laboratories.

Other lesser-used types of PPE that may need to be considered include ear protection (plugs and defenders), footwear (suitable toe protection and no open-toe sandals), aprons (for emptying autoclave tins), plasters (applied before entering laboratory over any breaks in the skin), and finger cots where sharps are being used.

5.4 Use of PPE

Before using any forms of PPE, workers must be instructed and trained in its correct use as well as checking PPE before use. Any defects in PPE should be reported before removal from use. In summary, PPE needs to be the following:

- appropriate
- complaint with legal standards (CE marking)
- ergonomic
- correctly fitting
- compatible with other forms of PPE used
- compatible with substances being used
- maintained
- having specified storage
- assessed before each use.

Any PPE that may be contaminated by biological agents must be removed on leaving the working area and kept apart from uncontaminated clothing and equipment. Any PPE that is contaminated by biological agents must be either decontaminated and cleaned or, if necessary, destroyed as it may be classed as "hazardous waste."

Summary

PPE should be considered as a last resort when undertaking a risk assessment. When all other options have been discarded, then PPE should be considered. Workers need to be appropriately trained in its use as well as any preuse checks. Remember, PPE only protects the wearer and if used incorrectly the protection afforded may be considerably reduced.

6
Shipping of Infectious Substances According IATA-DGR Regulations

Mandy Elschner and Martin Heller

6.1
Introduction

All microbiological laboratories send and receive samples and are subject to regulations on packaging and labeling of shipments. Depending on whether shipment is by road, rail, aircraft, or by ship, regulations can differ in some key points.

This chapter gives an overview about the most important points of the International Air Transport Association-Dangerous Goods Regulations (IATA-DGR) [1]. The IATA-DGR are updated and published every year on the basis of the agreement of the UN Committee of Experts, the International Atomic Energy Agency, and the International Civil Aviation Organization (ICAO).

The regulations for transport on the road are given in the ADR [2] (Accord européen relatif au transport international des marchandises dangereuses par route), for transport by rail in the RID (Règlement concernant le transport international ferroviaire de marchandises dangereuses) and both are based on the IATA-DGR.

6.2
Classifications and UN Code

All dangerous goods (DGs) are divided into different classes as shown in Table 6.1.

DGs in class 6.2 are infectious substances and could be any of the following (Table 6.2):

- infectious substances
- genetically modified microorganisms and organisms
- cultures
- biological products
- patient specimens
- medical or clinical waste.

Working in Biosafety Level 3 and 4 Laboratories: A Practical Introduction, First Edition.
Edited by Manfred Weidmann, Nigel Silman, Patrick Butaye, and Mandy Elschner.

Table 6.1 Dangerous goods classification.

Class 1	Explosives
Class 2	Gases
Class 3	Flammable liquids
Class 4	Flammable solids
Class 5	Oxidizing substances and organic peroxide
Class 6	Toxic and infectious substances
Division 6.1	Toxic substances
Division 6.2	Infectious substances
Class 7	Radioactive material
Class 8	Corrosives
Class 9	Miscellaneous dangerous goods (dry ice)

Table 6.2 The infectious substances must be assigned to a UN category.

Infectious substances	UN code
Infectious substances (category A)	UN 2814
	UN 2900
Biological substance (category B)	UN 3373
Genetically modified microorganisms and organisms	UN 3245
Medical or clinical waste	UN 3291

In order to determine the right classification, the following definitions should be considered:

Infectious substances. Infectious substances are defined as substances that are known or are reasonably expected to contain pathogens. *Pathogens* are defined as microorganisms (including bacteria, viruses, rickettsiae, parasites, and fungi) and other agents such as prions that can cause disease in humans or animals.

Cultures. Cultures are the result of a process by which pathogens are intentionally propagated.

Patient specimens. Patient specimens are human or animal materials, collected directly from humans or animals, including, but not limited to, excreta, secreta, blood and its components, tissue and tissue fluid swabs, and body parts being transported for purposes such as research, diagnosis, investigational activities, disease treatment, and prevention.

Biological products. Biological products are products from living organisms that are manufactured and distributed in accordance with the requirements of appropriate national authorities and are used for prevention, treatment, or diagnosis of diseases. They include, but are not limited to, finished or unfinished products such as vaccines.

Category A: It refers to an infectious substance that is transported in a form that, when exposure to it occurs, is capable of causing permanent disability, life-threatening disease, or fatal disease in otherwise healthy humans or animals. For these, UN 2814 or UN 2900 will apply depending on whether they are specific to humans or animals, respectively, or both.

The proper shipping name for UN 2814 is "Infectious substance, affecting humans."

The proper shipping name for UN 2900 is "Infectious substance, affecting animals."

The microorganisms that are listed as category A are published in the IATA-DGR, and examples are given in Table 6.3.

Category B: It refers to an infectious substance that does not meet the criteria for inclusion into category A. Infectious substances in category B must be assigned to UN 3373.

The proper shipping name for UN 3373 is "Biological substance, Category B."

If a laboratory wants to ship an isolate of *Bacillus anthracis*, the proper shipping name would be "Infectious substance, affecting humans" and the appropriate UN code would be UN 2814. In the case of a blood sample from an animal (no culture), the proper shipping name is "Biological substance, Category B" and the UN code is UN 3373. Exceptions are possible in the case of samples with a minimal likelihood of pathogens being present. For example, samples for blood or urine tests to monitor hormone, blood glucose levels, or blood products for transfusions can be sent as "Exempt human specimen" or "Exempt animal specimen." Animal carcasses infected with category A pathogens have to be classified as category A with the UN code UN 2814 or UN 2900.

6.3 Limitations

It is prohibited to carry DGs falling within class 6.2 by vehicles carrying passenger or crew or to send DGs in class 6.2 by airmail.

On the road DG in class 6.2 with UN codes, UN 2814 and UN 2900 have to be transported by a trained driver in a specially equipped and labeled vehicle. Biological substances falling within category B (UN 3373) can be transported by a private person in a private car.

6.4 Packaging

After finishing the classification of the shipment, the proper packaging material has to be chosen. The IATA-DGR award different UN numbers to specific packing instructions (PIs).

The three PIs that concern category A and category B shipments are:

- PI 650: UN 3373
- PI 620: UN 2814 + UN 2900
- PI 904: UN 1845 (carbon dioxide, solid, dry ice).

Table 6.3 Indicative examples of infectious substances included in category A (Copyright @ United Nations, 2011, reprinted from [3]).

Indicative examples of infectious substances included in category A in any form unless otherwise indicated (this list is not exhaustive)

UN-No proper shipping name	Microorganism	
UN 2814 Infectious substance, affecting humans	*Bacillus anthracis* (cultures only)	Highly pathogenic avian influenza virus (cultures only)
	Brucella abortus (cultures only)	Japanese encephalitis virus (cultures only)
	Brucella melitensis (cultures only)	Junin virus
	Brucella suis (cultures only)	Kyasanur forest disease virus
	Burkholderia mallei–Pseudomonas mallei–Glanders (cultures only)	Lassa virus
	Burkholderia pseudonallei–Pseudomonas pseudomallei (cultures only)	Machupo virus
	Chlamydia psittaci–avian strains (cultures only)	Marburg virus
	Clostridium botulinum (cultures only)	Monkeypox virus
	Coccidioides immitis (cultures only)	*Mycobacterium tuberculosis* (cultures only)
	Coxiella burnetii (cultures only)	Nipah virus
	Crimean-Congo hemorrhagic fever virus	Omsk hemorrhagic fever virus
	Dengue virus (cultures only)	Poliovirus (cultures only)
	Eastern equine encephalitis virus (cultures only)	Rabies virus (cultures only)
	Escherichia coli, verotoxigenic (cultures only)	*Rickettsia prowazekii* (cultures only)
	Ebola virus	*Rickettsia rickettsii* (cultures only)
	Flexal virus	Rift Valley fever virus (cultures only)
	Francisella tularensis (cultures only)	Russian spring–summer encephalitis virus (cultures only)
	Guanarito virus	Sabia virus
	Hantaan virus	*Shigella dysenteriae* type 1 (cultures only)
	Hantavirus causing hemorrhagic fever with renal syndrome	Tick-borne encephalitis virus (cultures only)
	Hendra virus	Variola virus
	Hepatitis B virus (cultures only)	West Nile virus (cultures only)
	Herpes B virus (cultures only)	Yellow fever virus (cultures only)
	Human immunodeficiency virus (cultures only)	*Yersinia pestis* (cultures only)

Table 6.3 (*Continued*)

Indicative examples of infectious substances included in category A in any form unless otherwise indicated (this list is not exhaustive)

UN-No proper shipping name	Microorganism
UN 2900 Infectious substance, affecting animals	African swine fever virus (cultures only) Avian paramyxovirus type 1 velogenic Newcastle disease virus (cultures only) Classical swine fever virus (cultures only) Foot-and-mouth disease virus (cultures only) Lumpy skin disease virus (cultures only) *Mycoplasma mycoides* – contagious bovine pleuroneumonia (cultures only) Pestes des petits ruminants virus (cultures only) Rinderpest virus (cultures only) Sheep-pox virus (cultures only) Goatpox virus (cultures only) Swine vesicular disease virus (cultures only) Vesicular stomatitis virus (cultures only)

6.5 Packing Instruction 650 for Biological Substance, Category B

The packaging must be of good quality, strong enough to endure shocks during loading and unloading usually encountered during transport. The packaging must consist of three components: (i) primary receptacle(s), (ii) secondary packaging, and (iii) rigid outer packaging (Figure 6.1).

Primary receptacles must be packed into secondary packaging in such a way that, under normal conditions of transport, they cannot break, be punctured, or leak their contents into the secondary packaging. Secondary packaging must be secured in the additional outer packaging with suitable cushioning material. Any leakage of the contents must not compromise the integrity of the cushioning material or the outer packaging.

For liquid and solid substances, the primary receptacle(s) must be leakproof/siftproof and must not contain more than 1 L and must not exceed the outer packaging weight limit of 4L/4kg. The secondary packaging must be leakproof/siftproof.

If multiple fragile primary receptacles are placed in a single secondary packaging, they must be either individually wrapped or separated from each other to prevent contact between them. Absorbent material must be placed between the primary receptacle and the secondary packaging. The absorbent material, such as cotton wool, must be in sufficient quantity to be able to absorb the entire contents of the primary receptacle(s) so that any release of the liquid substance will not compromise the integrity of the cushioning material or the outer packaging. The

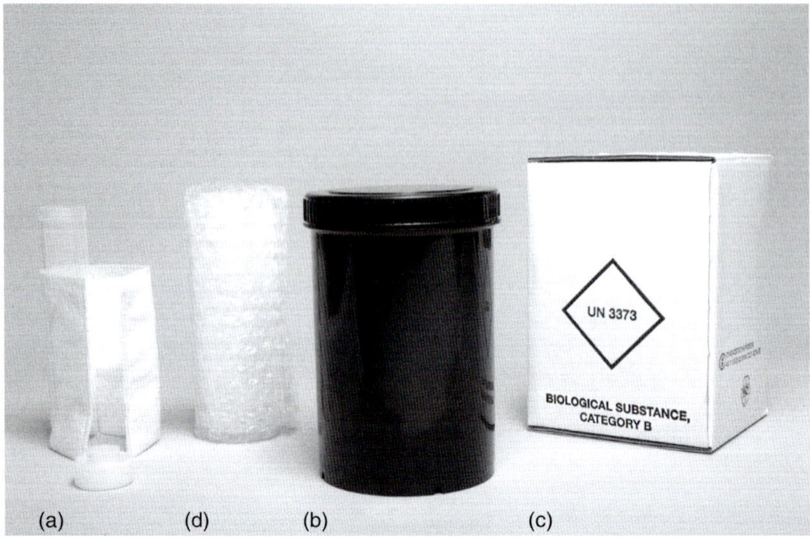

Figure 6.1 Package for biological substances, category B (UN 3373): (a) primary receptacle with absorbent material, (b) secondary packaging, (c) rigid outer packaging, and (d) cushioning material.

primary receptacle or the secondary packaging must be capable of withstanding, without leakage, an internal pressure of 95 kPa in a range of +40 to 55 °C (−40 to 130 °F). An itemized list of contents must be enclosed between the secondary packaging and the outer packaging. At least one surface of the outer packaging must have a minimum dimension of 100 mm (4 × 4 in.). Packages containing biological substances must be clearly marked "Biological Substance, Category B." Packages must be marked with a diamond-shaped symbol with the characters "UN 3373" with a minimum size of 50 × 50 mm or minimum 2 × 2 in. as shown in Figure 6.2.

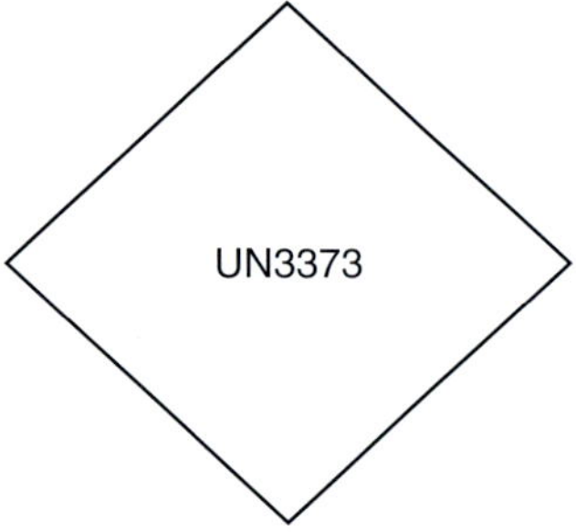

Figure 6.2 Label for infectious substance, catergory B

The list of markings and labels for category B packages includes the following:

- Markings:
  - shipper's name, address, and telephone number
  - receiver's name, address, and telephone number
  - UN number
  - proper shipping name.
- Labels:
  - UN 3373 label.

The completed package must be capable of successfully passing the drop test from a height of no <1.2 m. This is tested by the manufacturers of the commercially available packing materials, and there is no requirement to test it again by the shipper. All markings must be clearly visible, and in the case of using an overpack, the package markings must be reproduced on the outside and the overpack must be marked with the word "Overpack." A shipper's declaration for DGs is not required.

6.6 Packing Instruction 620 for Infectious Substance, Category A; UN 2814 and UN 2900

Shippers of infectious substances must comply with these regulations and must ensure that packages are prepared in such a manner that they arrive at their destination in good condition and present no hazard to persons or animals during transport.

The packaging must consist of three components:

1) inner packaging, consisting of:
   - watertight primary receptacle(s);
   - watertight secondary packaging;
   - absorbent material in sufficient quantity to absorb the entire contents placed between the primary receptacle and the secondary packaging.
2) an itemized list of contents must be enclosed between the secondary packaging and the outer packaging.
3) a rigid outer packaging, with a smallest external dimension of 100 mm.

The packages must be tested, for example, by the drop test procedure, or for internal pressure resistance in order to pass and receive UN-specification markings.

UN-specification markings:

Packages containing infectious substances, category A must be marked with a diamond-shaped label with the characters "Infectious Substance" and have a minimum size of 50 × 50 mm or minimum 2 × 2 in. as shown in Figure 6.3.

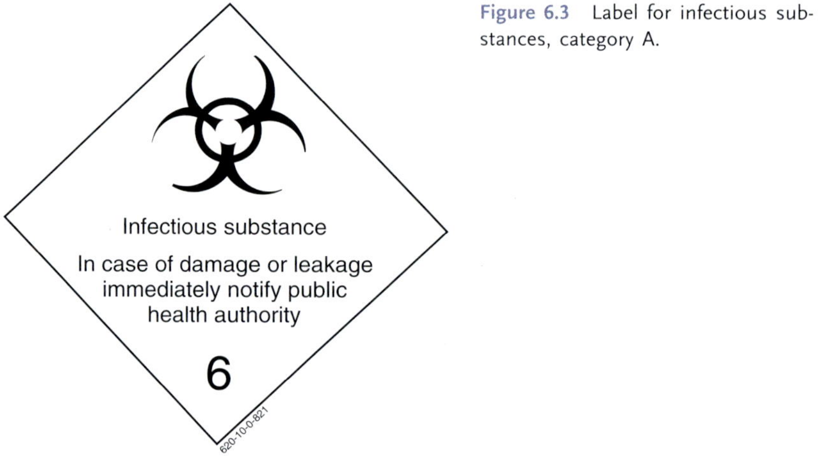

Figure 6.3 Label for infectious substances, category A.

The list of markings and labels for category A packages includes the following:

- Markings:
  - shipper's name, address, and telephone number;
  - receiver's name, address, and telephone number;
  - name and telephone number of responsible person (available 24 h until shipment arrives);
  - UN specification marking;
  - proper shipping name and UN number.
- Labels:
  - infectious substance label (minimum size 50 × 50 mm or minimum 2 × 2 in.)
  - package orientation label (only used when primary container exceeds 50 ml) (Figure 6.4).

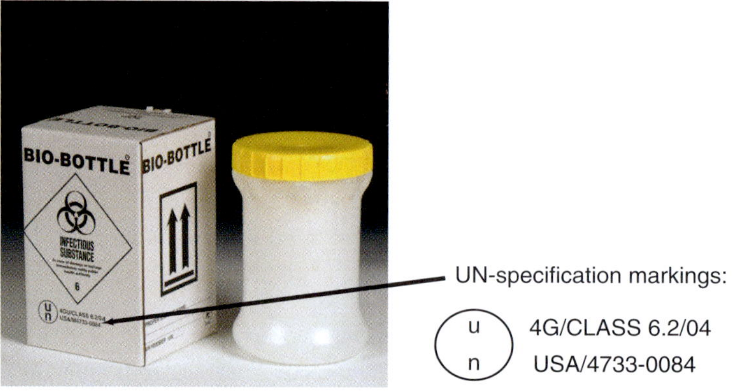

Figure 6.4 Package for infectious substances, category A (UN 2814 and UN 2900).

Figure 6.5 Overpack used for infectious substances, category A and dry ice.

Unless all package markings are clearly visible, the following requirements apply to the overpack. It must be marked with the word "Overpack," and the package markings must be reproduced on the outside of the overpack as shown in Figure 6.5.

Substances shipped refrigerated or frozen [wet ice, cooler packs, and carbon dioxide solid (dry ice)]: ice, carbon dioxide, solid (dry ice), or other refrigerant must be placed outside the secondary packaging(s) or alternatively in an overpack with one or more complete packages. Interior support must be provided to secure the secondary packaging(s) in the original position after the ice or carbon dioxide, solid (dry ice) has been filled. If ice is used, the packaging must be leakproof. If solid carbon dioxide (dry ice) is used, the outer packaging must permit the release of carbon dioxide gas. The primary receptacle and the secondary packaging must maintain their containment integrity at the temperature of the refrigerant used as well as at the temperatures and pressure(s) of air transport to which the receptacle could be subjected if refrigeration were to be lost.

6.7 Packing Instruction 904 (UN 1845) for Dry Ice

Solid carbon dioxide (dry ice), when used for transport by air, must be in packaging designed and constructed to permit the release of carbon dioxide gas and to prevent a buildup of pressure that could rupture the packaging. The UN code "UN 1845" and net weight in kilogram (kg) of the solid carbon dioxide (dry ice) and the class 9 label (Figure 6.6) must be marked on the outside of the package. Arrangements between shipper and operators must be made for each shipment, to ensure that ventilation safety procedures are followed. The shipper's declaration requirements are only applicable when the solid carbon dioxide (dry ice) is used as a refrigerant for

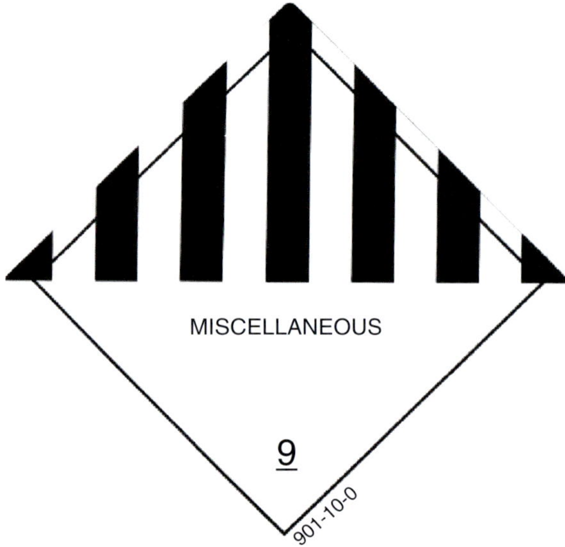

Figure 6.6 Label for class 9 dangerous goods as dry ice.

DGs that require a shipper's declaration (UN 2814 and UN 2900). When a shipper's declaration is not required, the information required for packages shipped using dry ice must be contained in the "Nature and Quantity of Goods" box on the air waybill, excluding the PI number 904 and packing group III.

6.8 Documentation

For each UN 3373-, UN 2814-, or UN 2900-package, an air waybill of the respective company has to be filled out for the transportation stage on the road en route to the airport. For packages containing material with UN code 2814 or 2900, in the column detailing handling information, the statement "Dangerous Goods as per attached Shipper's Declaration" has to be made and a shipper's declaration must be prepared. This shipper's declaration has to be filled out according to IATA-DGR regulations and has to be signed by a trained shipper as shown in the example in Figure 6.7. The shipper is responsible for the correct classification, packaging, and labeling of the package and needs to be retrained including an examination every 2 years.

SHIPPER'S DECLARATION FOR DANGEROUS GOODS

Shipper
Dr. Sabine Mustermann
Institute of Microbiology
Mustermannstrasse 1
10115 Berlin, Germany
phone: +49 1234 56789

Air Waybill No
Page 1 of 1 Pages
Shippers Reference Number (optional)

Consignee
Dr. Robert Mustermeier
Institute of Virology
Mustermeierstrasse 2
20095 Hamburg, Germany
phone: +49 3456 78900

WORLD COURIER

Two completed and signed copies of this Declaration must be handed to the operator

TRANSPORT DETAILS

This shipment is within the limitations prescribed for: (delete non-applicable)
PASSENGER AND CARGO AIRCRAFT | CARGO AIRCRAFT ONLY XXXX

Airport of Departure

Airport of Destination:

WARNING

Failure to comply in all respects with the applicable Dangerous Goods Regulations may be in breach of the applicable law subject to legal penalties.

Shipment type: (delete non-applicable)
NON-RADIOACTIVE | XXXXXXXXX

NATURE AND QUANTITY OF DANGEROUS GOODS

Dangerous Good Identification						
UN or ID No.	Proper Shipping Name	Class or Division (Subsidiary Risk)	Pack-ing Group	Quantity and type of packing	Packing Instr.	Authorization
UN2814	Infectious substance, effecting humans (Ebola) (liquid)	6.2		1 Fibreboard box x 20ml	620	
UN1845	Dry Ice	9	III	8 kg	904	
				Overpack used		

Additional Handling Information

24 Hour emergency Contact Number: Dr. Sabine Mustermann phone: +49 1234 56789

I hereby declare that the contents of this consignment are fully and accurately described above by the proper shipping name, and are classified, packaged, marked and labelled/placarded, and are in all respects in proper condition for transport according to applicable international and national governmental regulations. I declare that all of the applicable air transport requirements have been met.

Name/Title of Signatory
Dr. Sabine Mustermann Microbiologist
Place and Date
Berlin, Germany, 01.02.2012
Signature (see warning above)
Mustermann

Figure 6.7 Shipper's declaration.

Summary

The regulations for transport of infectious substances via aircrafts are regulated by the "IATA" and published annually in the "IATA-DGR." In this chapter, the detailed instructions of the IATA-DGR are presented in a simple and clear manner. Infectious substances and related biological materials belong to the DGs class 6.2, consisting of infectious substances (category A, UN 2814, UN 2900), biological substances (category B, UN 3373), genetically modified microorganisms and organisms (UN 3245), and medical or clinical waste (UN 3291). The IATA-DGR award different UN code numbers to specific PIs: PI 650 to UN 3373, PI 620 to UN 2814 and UN 2900, and PI 904 to UN 1845 (dry ice). The PIs include all prescribed specifications for the packing material, markings, and labeling. For materials of category A and category B, an air waybill is necessary, and for category A material, a shipper's declaration according to IATA-DGR regulations has to be filled out and signed by an authorized shipper. The shipper is responsible for the accurate classification, proper packaging, and labeling of the package and needs training including an examination every 2 years.

References

1. International Air Transport Association (IATA) (2011) *IATA Dangerous Goods Regulations Manual 2011 (DGR)* 52th edn 2010, IATA Effective: 1 January–31 December.
2. Conrad, J. (2011) *Gefahrgutrecht ADR/RID*, WEKA MEDIA GmbH & Co.KG, Kissing.
3. United Nations (2011) UN *Recommendation on the Transport of Dangerous Goods (Model Regulations)*, 17th revised edition.

7
Disinfection and Decontamination

Patrick Butaye

7.1
Introduction

When working in a BSL-3 (biosafety level) or BSL-4 laboratory, one needs to make sure that the workspace is not infectious for others or oneself. As one can never be absolutely certain that nothing has been spilled, environmental contamination should always be assumed when finishing work.

There is a major difference between disinfection and decontamination. While the first is aiming at reducing the microbial load to avoid infections, the latter is aiming at completely killing all microbes. However, even then, this is only a relative thing as the detection of live microbes has its limits and is dependent on the method used and also on the microbe concerned. Some microbes (e.g., easily cultivable *Bacillus anthracis*) are detectable at far lower levels than, for example, viruses that have to be cultivated in cell cultures. Moreover, cell cultures are also susceptible to inactivation by the disinfectant used, and this may influence cytopathogenic effects.

Apart from potential contamination of surfaces, one should also suspect air to be contaminated. Manipulations such as centrifugation may create aerosols, which may contaminate unexpected surfaces and infect people in the laboratory. It is clear that specific measures should be taken to avoid generation of aerosols; however, in case, they have been generated, all the surfaces and the air in the laboratory should be decontaminated. There are many different surfaces with their own characteristics. Therefore, decontamination of surfaces requires specific attention.

There are several ways of performing disinfection/decontamination, depending on the nature of the material or environment, which needs to be disinfected/decontaminated. In this chapter, the different ways of disinfection and decontamination and the influencing factors are discussed.

Commercially available products for disinfection and decontamination have to undergo registration as regulated by the European Normation organization. There are strict rules on the performance characteristics of decontamination products. These performance characteristics are, however, not always adapted to the requirements of BSL-3/BSL-4 environments. The procedures for testing disinfectants and the influencing factors will be discussed. Finally, we will give

Working in Biosafety Level 3 and 4 Laboratories: A Practical Introduction, First Edition.
Edited by Manfred Weidmann, Nigel Silman, Patrick Butaye, and Mandy Elschner.

an overview of some of the compounds currently used for disinfection in the laboratory environment.

7.2 Ways of Decontamination/Disinfection

Different environments and equipment may need decontamination. As for equipment, different ways of decontamination are possible (Figure 7.1). The permanent or disposable nature of equipment is of importance for the choice between physical or chemical decontamination. Physical decontamination often destroys the material, which then turns into waste. When the material is nondisposable or when it is valuable, other means can be used. Depending on the nature of the material, different methods are available. If the nature of the equipment, for example, cannot withstand liquids, then gas, temperature-based, or irradiation-based decontamination can be used. If the material cannot withstand high temperature, then gas or irradiation is possible. If the material cannot withstand the activity of fumigants, as, for example, some electronic or optical materials, irradiation may be opportune. The access to the latter is of course quite limited and is an expensive method. Therefore, the financial aspect plays a role when deciding on a decontamination strategy. In Figure 7.1, a summary decision scheme for decontamination of equipment is shown.

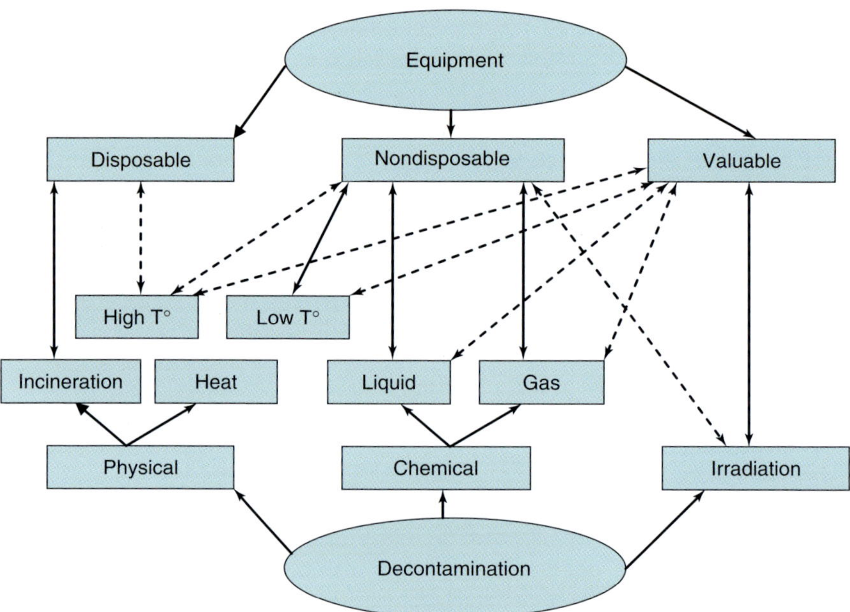

Figure 7.1 Decontamination decision tree.

As to environmental decontamination (worktables, rooms, and centrifuges), a similar decision scheme is followed. Of course when talking about rooms, few options will be left. First, the physical methods cannot be used. Flooding a place up to the ceiling is impossible, and irradiation is not feasible for a room. Therefore, only decontamination by gas (fumigation) remains. This is discussed in Chapter 8 of this book.

7.3 Physical Disinfection/Decontamination

Physical decontamination only encompasses heat inactivation. The most drastic method is incineration, which completely destroys all organisms. However, an incinerator, in close contact to a BSL-3/BSL-4 facility, is first expensive and second often of low feasibility because of the size and the demands of an incinerator. It however may be employed in an ABSL-3 (animal biosafety level) facility as described in Chapter 4.

A second method is autoclaving with or without the use of steam. The regular autoclaving process uses steam at a temperature of 121 °C under a pressure of $1\,\mathrm{kg\,cm^{-2}}$ for 15–20 min. Changes in temperature may decrease the time necessary for autoclaving. More thorough autoclaving is achieved by increasing either temperature or time. In particular, for spore-forming bacteria, 135 °C for 30 min is recommended. The relevant temperature measured in the autoclave is the actual temperature of the material that needs to be decontaminated. Therefore, when working with infectious materials, run times may be prolonged up to 60 min.

Decontamination without the use of steam needs far higher temperatures. Autoclaving denatures proteins; denaturation of proteins without the use of moisture requires a far higher temperature, which is mostly 160 °C. This temperature needs to be held for at least 2 h. One can shorten the time to 1 h at 180 °C. This is sometimes referred to as *sterilization*.

An exception is the decontamination of prions. These are more resistant than any of the other infectious agents. Autoclaving needs to be prolonged to 1–1.5 h at a temperature between 132 and 134 °C and a pressure of $2\,\mathrm{kg\,cm^{-2}}$.

Decontamination by filtration is used for decontamination of outgoing air and sometimes the incoming air in BSL-3/BSL-4 laboratories (Chapter 3).

7.4 Irradiation

Irradiation can be done with UV radiation or ionizing radiation. UV radiation acts on cellular nucleic acids. The wavelength used for microorganisms is 254 nm. It is used to decontaminate surfaces and air. It should be noted that there is no penetration through organic material, water, and so on; therefore, the surface must be well cleaned and dried before UV radiation can be effective. Other influencing factors are the intensity of the light (which diminishes with the lifespan of the lamp), distance between the lamp and the exposed bacteria, relative humidity, and bacterial species. In contrast to widespread practice, it therefore does not make sense to use UV irradiation in biosafety cabinets (BSCs) for extended time periods (e.g., overnight).

Ionizing radiation, such as γ-radiation, is not frequently used to decontaminate materials from BSL-3/BSL-4 laboratories. However, it is used in some cases for valuable materials. It is clear that such an installation is rarely in the close vicinity of the laboratory. Therefore, material needs to be packed and the airtight package should be externally decontaminated in order to bring it safely to the irradiation plant.

7.5
Factors Influencing Chemical Disinfection/Decontamination

Several chemicals are used to disinfect and decontaminate. Each has its own characteristic efficacy and shows optimal action under specific conditions. The inappropriate use of disinfection/decontamination products may hamper the activity and may result in an incomplete disinfection/decontamination. Knowledge of the factors influencing the activity of a product is a first step in a good disinfection/decontamination protocol.

7.5.1
Temperature

Temperature may influence the disinfecting activity of chemicals in two ways. First, it may act directly on the chemical itself and second, it may alter the metabolic activity of the organism to be disinfected. For some disinfection protocols, as, for example, formaldehyde fumigation, a minimum temperature is required. Below 7 °C, formaldehyde fumigation is ineffective. Most disinfectants act on nongrowing bacteria (contrary to antibiotics that need growing bacteria to be active). However, the activity of the disinfectant increases in general when bacteria are in the logarithmic growing phase rather that in the stationary phase. In general, an increased temperature will increase the activity of disinfectants.

7.5.2
Time of Contact

Time of contact may have an influence on the activity of a disinfectant. In the case of heat inactivation, this depends largely on the temperature used. Some disinfectants may be quite corrosive and for this reason, contact time may be limited in order not to deteriorate the surfaces or materials to which it is applied.

7.5.3 Microorganism

Not all microorganisms may be as susceptible to inactivation as others. Some have developed specific resistance mechanisms for their habitat. They may resist high temperatures (as, e.g., microorganisms living in hot water). Some have developed specific strategies to resist temporarily unfavorable circumstances, as exemplified by the sporulation of certain bacteria.

In addition, some bacteria may have acquired resistance mechanisms, as exemplified by the quaternary ammonium compound (QAC) efflux pump of Gram-negative bacteria integrated in the 5′ end of integrons, which frequently are located on mobile genetic elements such as plasmids and can move from one bacterium to another. This efflux pump causes resistance against quaternary ammonium disinfectants.

The hierarchy of resistance to chemical treatment beyond the prions is lead by spore-forming bacteria, such as *Bacillus* spp. and *Clostridium* spp., followed by mycobacteria, both from the tuberculosis group and from the non-tuberculosis group. Next in line are nonenveloped small viruses including parvovirus, poliovirus, coxsackievirus, and rhinovirus; fungi, including yeasts; and then Gram-positive and Gram-negative vegetative bacteria. The latter two are the largest group of bacteria including Enterobacteriaceae, staphylococci, enterococci, and most of the BSL-3 bacteria. Last are the enveloped viruses such as influenza or Ebola virus. An exception among the enveloped viruses is the severe acute respiratory syndrome coronavirus (SARS-CoV), which features tightly packed proteins, which makes it as resistant as nonenveloped viruses [1, 2].

7.5.4 Surface Type (Absorbant vs Nonabsorbant)

Some surfaces are easier to disinfect than others because of their physicochemical characteristics. In general, laboratory surfaces can be easily cleaned and disinfected. However, some surfaces may be absorbent. Good examples are the interior of fridges and laboratory gowns. When spills happen, one should have an appropriate disinfection procedure for these surfaces but testing protocols for these surfaces is generally not done.

7.5.5 Liquid

The presence of liquids may have a dual effect on the decontamination procedure. First, it may dilute the disinfectant to a suboptimal concentration. Second, the liquid may contain other compounds such as organic material, which inactivate the disinfectant, or compounds that may react with the disinfectant to produce secondary products.

7.5.6
pH

Some pH conditions are unfavorable for the action of certain chemicals against microorganisms. As an example, the combination of a liquid with a high pH with bleach (NaOCl) may cause the production and release of chlorine gas.

7.5.7
Presence and Type of Dirt

In general, dirt is a complex chemical mixture that may inactivate the disinfectant. It also increases the volume to be disinfected, and in a similar way, there is a diluting effect of the chemical. The presence of dirt requires that the chemical disinfectant is applied for a longer time. Dirt and its content of fatty acids or proteins have an effect on disinfection and make them difficult to estimate whether disinfection has been successful. Therefore, cleaning should precede disinfection.

7.5.8
Concentration of the Product

Disinfectants are frequently corrosive and caustic. Using an optimal concentration is therefore of importance. A higher concentration is not necessarily more effective. It can even be less effective as shown by the best disinfectant activity of alcohols at 70% – higher concentrations tend to be less effective. Optimal concentrations are generally given for clean areas. Dirt on the disinfecting surfaces may need a higher concentration of the disinfectant.

7.5.9
High-Pressure Water Cleaning

Different surfaces with different structures may need a different way of cleaning. In general, laboratory materials and laboratory equipment are made to be easily cleaned and disinfected. When looking at experimental animal facilities, the situation is quite different. For example, the floor may be soft or have a profile to stop animals from slipping. The use of high water pressure with a detergent may be appropriate to clean those; however, it causes aerosols and may be dangerous. In these environments, one should also be aware that the floor may contain cracks and other lesions, which are difficult to clean. Using water at high pressure for a longer time leads to deeper penetration and cleaning or disinfection. However, it also increases the volume of waste, and an optimal use of water is of economic importance. Note also that the dispersion of small quantities of fluid disinfectant over a larger surface may be quite difficult and care should be taken that the whole surface is adequately disinfected.

7.5.10 Water Used

In different geographical localizations, different components are added to tap water to assure its quality. NaOCl is added in many countries, although there are no large differences in concentrations added. Calcium/magnesium concentration defined as the hardness of the water may differ largely from area to area. Other contents of water (minerals, proteins, glycides, and lipids) may also influence the activity or even inhibit some disinfectants.

7.5.11 Mechanism/Methods of Decontamination

Methods of decontamination are different according to the products used. Some are used as liquids, some as solid materials, while others are used as a gas or are nebulized. Decontamination by gas or nebulization allows fumigant to reach the small corners of an environment, whereas, for example, the use of a solid disinfectant such as CaO will only allow the disinfection of more or less horizontal surfaces. When using liquids for decontamination, the contact time for vertical surfaces needs to be increased. This will also depend on the humidity of the environment.

7.5.12 Inoculum Concentration

Killing curves for disinfection are not linear. Therefore, the disinfectant concentration will determine the success of reducing the number of bacteria. There may also be a lag of the regrowth after bacteria have been exposed to a disinfectant, which affects growth control checks. The latter depends on the disinfectant used.

7.6 Testing the Activity of a Certain Product

As stated earlier, the activity of disinfectants is influenced by many conditions. Therefore, standardized protocols are developed to assess the effectiveness of chemical disinfectants and for the type of their physical application.

7.6.1 Physical Disinfection

Physical disinfection control is mainly based on the capacity of the method to inactivate control organisms. In general, the most resistant organisms are used. For temperature-based methods, the temperature-resistant and easily cultivatable *Bacillus sporothermodurans, Bacillus subtilis, or Bacillus stearothermophilus* are used.

These agents can be bought in vials or as spore strips to place them in an autoclave, for example.

7.6.2 Chemical Disinfection

7.6.2.1 Introduction

Testing of the effectiveness of chemical disinfection is subject to quite a lot of parameters as their usage in different situations is more difficult to assess. Standardized protocols are used to allow comparison of different chemicals. These protocols are more or less adapted to their use. For example, the ones used in industry or food production do not have the same requirements as those for the use in hospitals or animal rearing facilities. While disinfection activity is sufficient in one environmental decontaminating activity maybe needed in another.

In a BSL-3 or BSL-4 environment, there is a need for complete decontamination, as the consequences of human infection can be disastrous, or even fatal. When starting testing a product, it is advisable to test it according to the homologation procedures in order to compare different compounds. Standardized homologation consists of three phases.

7.6.2.2 Phase 1 Studies

In the *first phase*, the activity of the product is assessed in a standardized way by an *in vitro* test. They are quantitative suspension tests. Different forms of activity are assessed:

- bacteria: bactericidal–bacteriostatic
- fungi: fungicidal–fungistatic
- yeast
- mycobacteria: mycobactericidal/tuberculocidal
- spores (from bacteria): cidal/static
- viruses.

For bacteria, the following reference organisms are recommended: *Pseudomonas aeruginosa*, *Escherichia coli*, *Staphylococcus aureus*, *Enterococcus hirae*, and sometimes additionally *Salmonella typhimurium*, *Lactobacillus brevis*, and *Enterobacter cloacae*.

The homologation of a disinfectant requires a reduction of 5 $\log_{10}$ of the organisms. This renders a surface certainly not sterile but may lower the infectivity substantially. When working with highly pathogenic microorganisms, this may however be not sufficient.

In the case of sporicidal testing, a *Bacillus cereus* strain is proposed. This species is closely related to *B. anthracis* and can as such be used for *in vitro* testing [3]. However, it is also advisable to test first, more than one strain and second, the species that is actually being worked on.

As for viruses, the nonenveloped bovine parvovirus (known be very resistant to numerous disinfectants), poliovirus, and adenovirus are advised. However, the specific viruses to be treated should be tested. For the fungicidal activity, specific organisms listed are *Candida albicans* and *Aspergillus niger*. For more extensive testing *Saccharomyces cerevisiae, S. cerevisiae* var. diastaticus are recommended. Surprisingly, there is no test for basic viricidal activity. These tests are linked to different environments and conditions and are directly included in phase 2 studies. Similarly, the sporicidal and mycobactericidal activities are only defined in phase 2 studies. For fungicidal testing, the methodology used is the same as for bacterial testing, however, here only a 4 $\log_{10}$ reduction is required. The method used defines the water, the media (and components), diluents, neutralizers, and rinsing liquids. It also defines the apparatus, concentrations of microorganism, temperature to work at, and the contact time between disinfectant and microorganism in the tube. The way to test is likewise well described and well defined and allows a comparison between two different chemical disinfectants.

7.6.2.3 Phase 2 Studies

In *phase 2 studies*, the claims of activity of the product toward various environments are identified and verified in a standardized way. Of course, the circumstances used are not necessarily adapted to the needs of the BSL-3/4 environments. The environments that are defined for homologation are first the medical area. There a subdivision is made for hygienic hand wash or rub, surgical hand wash or rub, surface disinfection under clean and dirty conditions, and finally instrument disinfection under clean or dirty conditions. A second environment is the veterinary area. Here, the following categories are defined: surface disinfection with high and low level of soiling on porous and nonporous surfaces, immersion of highly soiled objects, and finally hygienic hand wash. A third environment is disinfectants for use in food, industrial, domestic, and institutional areas. Here, a lot of subdivisions can be defined. From clean to dirty surfaces, hygienic hand wash and rub are used, for the use in breweries and beverage (including milk) factories and for the use in cosmetic manufacturing. Finally, there are products homologated without a specifically defined area of application. Here, only the phase 1 studies are necessary demanding defined temperatures (20 °C), contact times (60 min for non-spore-forming bacteria and yeast and fungi; 120 min for spore-forming bacteria), a 5 $\log_{10}$ reduction for non-spore-forming bacteria, and a 4 $\log_{10}$ reduction for other bacteria, yeasts, and fungi. For viruses, there has to be at least a 4 $\log_{10}$ reduction during a contact time of 60 min.

It is certain that cleaning the environment before disinfection is most favorable for having a most efficient decontamination; however, in certain circumstances, for example, in a laboratory accident, cleaning before disinfection is not advisable and the infectious agent should be killed as soon as possible during one application of the decontaminant. Therefore, it is advisable to test out some conditions with dirt that you may encounter in the laboratory. For example, when dealing with animal experiments, it is sometimes very difficult to clean before disinfection. A first decontamination under "dirty circumstances" may be necessary before the cleaning can be started. After this first decontamination and the cleaning, a more through disinfection can be performed. The phase 2 studies consist of two steps.

Phase 2, Step 1 Step 1: Quantitative suspension tests to establish that a product has bactericidal fungicidal or sporicidal or virucidal activity simulating practical conditions appropriate to its intended use.

1) In clean conditions, one starts from the viewpoint that nothing is really absolutely clean so that no inhibiting substances are present; therefore, a low concentration of bovine abumin is added (3 ppm). As for dirty conditions, bovine albumin is added at a concentration of 30 ppm. Testing both allows you to evaluate the inhibitory effect of dirt in general. Additional conditions may also be tested and should be adapted to the specific use of the disinfectant. For example, it is of interest to test the influence of tissue culture medium on the disinfection capacity in a virology laboratory. This would test the efficacy of your disinfection in case of an accident with a tissue culture flask.
2) Water hardness varies substantially between regions. Therefore, different types of water are tested including distilled water, and water differing in hardness.
3) Single components can also be tested for their inhibitory effect. This can be interesting for known dirt concentrations on surfaces to estimate the inhibitory effect of it on the disinfecting procedure. As for sugars, sucrose can be used. It is obligatory to test for sugars for homologation for the beverage and soft drink industries. In the cosmetic industry, sodium lauryl sulfate needs to be tested.
4) To simulate complex dirt, one may use milk. It contains sugars, proteins, and fats. Milk however is not a standardized substance and its contents may vary. Commercially available yeast extract, for the preparation of culture media, is much more homogeneous than milk. For elevated dirt conditions, as encountered when performing animal testing, a combination of yeast extract and albumin may be used.
5) In general, two additional pH ranges (neutral pH in phase 1), a slightly acidic pH (pH 5) and a basic pH (pH 9) need to be tested.

Phase 2, Step 2 In step 2, porous or nonporous surfaces are defined. This includes hygienic/surgical hand wash or rub and instruments used in the medical and veterinary areas.

7.6.2.4 Phase 3 Studies

Phase 3 studies are not performed for the homologation of a specific product by the producers and thus they aim to verify the claim of activity under specific conditions against specific agents. There are no defined or validated tests for this phase available. However, it is the most important phase for the work performed in the BSL-3/4 laboratories. To perform tests of decontamination that suit the laboratory practices, the requirements for the disinfectants need to be set by the user to allow effective decontamination of working bench, gloves, and materials. This is where evidence-based biosafety plays a role and the proof produced by the BSL-3/4 staff needs to be discussed with and accepted by biosafety officers.

7.7 Chemical Compounds Used as Disinfectants

7.7.1 Introduction

Many compounds are used for disinfection and they are frequently combined in disinfection products. In this section, the characteristics of compounds are described [4–6].

7.7.2 Phenols

Only phenolic derivatives are used, as phenol itself is quite toxic, carcinogenic, and corrosive. In some cases, skin irritation may still occur. A large variety of products are available most frequently containing *o*-phenylphenol, *o*-benzyl-*p*-chlorophenol, and *p*-*tert*-amylphenol.

Phenols are not used in combinations and at regular concentrations are not suitable for viruses (only enveloped lipophilic viruses), and not effective against bacterial spores. They may become more virucidal when a detergent is added or in some cases when the phenol concentration is increased. Although they are not really inactivated by organic materials, the addition of a detergent strongly improves the activity of the phenols. The phenols can also be incorporated into soaps and used for antisepsis but are unavailable in the European Union because of REACH (Registration, Evaluation, Authorisation, and restriction of CHemicals) legislation.

7.7.3 Chlorine Derivatives

Sodium hypochlorite (household bleach, NaOCl) is the most commonly used product. It is used at concentrations between 1 and 5% (and sometimes up to 10%). Although it is a very powerful disinfectant, its capacity strongly depends on the concentration used. Its activity is expressed as free chlorine, the active substance. As such, a 1/10 dilution of household bleach gives good disinfecting result for nearly every application. Two points are important to note when using bleach. First, bleach is very much inactivated by organic material, so cleaning is of utmost importance before disinfection. Second, its shelf life may decline rapidly after opening the container. Repeatedly, opening and closing a container during 1 month may result in a reduction of active chlorine of 50% [7].

Examples for other chlorine-based compounds are chlorine dioxide and inorganic and organic chloramines. Chlorine dioxide is a very reactive compound and active against spores. It can be used as a gas for fumigation of laboratories. Its use however is limited because of its toxicity, the difficulties to obtain it, as it needs to be generated at the application site, and its activity depends on relative air humidity (which should be between 70 and 80%). It is also used for decontamination of drinking water.

Inorganic chloramines are ammonium–chlorine compounds. These are mostly used for decontamination of potable water. Among the organic compounds, chloramine-T is the best known compound [8]. It is used for decontamination of surfaces and medical equipment, and is active also against spores. It can also be used for inactivation of dunk tanks in BSL-3 facilities.

7.7.4 Iodophores

Iodophores, for example, povidone–iodine, have a good and fast activity against bacteria, fungi, and viruses. They are however quite susceptible to inactivation by organic material and they may also cause skin irritation. These compounds are complexes that bind iodine, the active compound. They are however less toxic and cause less irritation than iodine. In general, iodophores are used for skin antisepsis and sometimes for disinfection of medical equipment. They are not suitable as laboratory surface disinfectants.

7.7.5 Quaternary Ammonium Compounds

A variety of these cationic surface detergents are used (alkyl benzalkonium chlorides, benzethonium chloride, cetrimide, and cetylpyrimidine chloride). They are not toxic and noncorrosive; however, they are less active on bacteria and show poor activity for nonenveloped viruses compared to other disinfectants. They are not particularly influenced by organic material. As such, they are mainly used for skin disinfection. Sometimes, they are also used for equipment disinfection, for example, when working with enveloped viruses.

In Gram-negative bacteria, an efflux pump causes resistance to quaternary ammonium compounds (Section 5.3). In Gram-positive bacteria, several efflux pumps, capable of causing resistance against quaternary ammonium compounds, have been described for *S. aureus* [9, 10].

7.7.6 Amphoteres

Amphoteres are surfactants or tension active agents. Nonionic detergents are also named amphoteres. They have an anionic radical and a cationic radical in their hydrophilic group. Some amphoteres have a strong antibacterial activity, for example, dodecylamino butyric acid and dodecyl-di-(aminoethyl) glycine. They are able to go through the bacterial membrane and enter into the cytoplasm of the bacteria. There they bind to dsDNA and inactivate the bacteria. Amphoteres are bactericidal and even mycobactericidal. They act however quite slow. They are neither toxic nor corrosive or caustic and little affected by organic material. As such, they are mainly used in food industry but can also be used for hand and skin disinfection.

7.7.7 Aldehydes

Several aldehydes such as formaldehyde, glutaraldehyde, *ortho*-phthalaldehyde are available for disinfection. Formaldehyde is mainly used as a gas and will be discussed in Chapter 8. The liquid form is rarely used. Glutaraldehyde is mainly used in its liquid form. It is noncorrosive. Its activity, which is at optimal conditions excellent and even includes bacterial spores, however depends on several factors, for example, pH, temperature, concentration, presence of inorganic ions, and age of the solution. The latter may be crucial because the shelf life is rather limited. It is almost not inactivated by organic material. Glutaraldehyde and formaldehyde readily evaporate and vapors (above 0.2 ppm) are irritating to eyes, nose, and throat and may also cause severe contact dermatitis, asthma, and rhinitis. Respiratory protection against vapors is advised. *Ortho*-phthalaldehyde does not vaporize so easily and is less susceptible to environmental conditions. Contact should be avoided at any time however, as it causes irritation. Its activity against spores is reduced. Unfortunately, resistance against these products has been reported for mycobacteria, methylobacterium, trichosporum, certain fungal spores, and cryptosporidium.

7.7.8 Calcium Oxide, Lime

Calcium oxide [CaO (s) + H_2O (l) $Ca(OH)_2$ (aqueous) ($\Delta Hr = -63.7\,kJ\,mol^{-1}$ of CaO)] or lime is a very active product that is little affected by organic material. It is however very corrosive toxic and caustic. It has long been proved to be excellent for environmental decontamination of contaminated fields. It is extensively used to decontaminate carcasses of cows killed by *B. anthracis*. The procedure includes an autopsy on the site of a suspect carcass to take samples, followed by on-site incineration of the carcass and the protective clothing. After the fire finishes, the ashes and surrounding ground are treated with lime and sprayed with water. The addition of water leads to a heavy reaction and calcium is all that remains. To this day, white layers of calcium can be found in the fields of Flandern from veterinary anthrax decontamination or treatment or mass graves of the great wars.

7.7.9 Alcohols

Ethanol and isopropanol are the two alcohols most extensively used for surface disinfection and antisepsis. Water enhances its activity and the optimal concentrations are between 60 and 90%. They have however a poor activity against bacterial spores, protozoan oocysts, and certain nonlipophilic (nonenveloped) viruses (e.g., adenoviruses). The major advantages are that their activity is rapid, they have a low toxicity, they do not irritate, and they are not caustic.

7.7.10
Chlorhexidine

Chlorhexidine is a noncaustic agent with low toxicity. Therefore, it is frequently used as wound disinfectant and antiseptic. It is also frequently used as surgical hand scrub. Its action is slower than alcohols. Its activity is much reduced by the presence of organic material. It is also incompatible with some detergents (e.g., anionic surfactants), so attention should be paid to the nature of the detergent used for cleansing.

7.7.11
Peroxides

Oxidative hydrogen peroxide vapors are frequently used in laboratories, fruit preservation, and aquaria. The major advantage of peroxides is their low toxicity for humans and animals, while it is sporicidal. Concentrations and incubation times for decontamination should be tested beforehand, because of many influencing factors, for example, relative humidity that should be below 30%. Peroxides are used for disinfecting wounds, indicating a low toxicity and no caustic or corrosive effects. The activity is related to the capacity of the organism to produce catalase (e.g., *S. aureus*). When contaminants with cellular catalase activity are suspected, a longer incubation (contact) time is needed. This is also the case for spores. Enhanced temperatures and concentration increase its activity. A major disadvantage is their fast inactivation by organic material. This renders these products only useful in a rather clean area. The activity of peroxides may be increased by the synergistic use of ultrasonic energy, UV radiation, and peracetic acid.

While more research is still needed, they can be used to reduce bacterial load in animal residencies, even when animals are present.

7.7.12
Peracetic Acid

Peracetic acid is an oxidative disinfectant, which is more potent than hydrogen peroxide in destructing spores and little influenced by the presence of organic material. It is however corrosive to several metals.

7.7.13
Sodium Hydroxide

This product at a concentration of 1N can be used to spot decontaminate spill sites before cleaning. This compound is however strongly irritating and corrosive. The surfaces one would want to disinfect should be resistant enough. It is also used as chemical decontaminant for liquids.

7.8 Conclusion

There are no rules for testing the efficacy of a certain disinfection/decontamination products, in a particular environment and conditions. It is therefore the task of the user to provide evidence-based data demonstrating the activity of the product for the application intended. The variety of constraints set by the materials and environment to be infected and the characteristics of the compounds to be tested demand a well thought through strategy and assessment before starting experiments. However, good proof of activity should convince biosafety officers and help to make life easy in BSL-3/4 laboratories.

Summary

Although used on a daily basis in BSL-3/4 laboratories, most laboratory staff generally know little about the ingredients of disinfection and decontamination agents. This chapter describes the homologation procedure used to validate these agents and gives a concise summary of the most common agents used.

References

1. Miriam, E.R. *et al.* (2004) Inactivation of the coronavirus that induces severe acute respiratory syndrome SARS-CoV. *J. Virol. Methods*, **121**, 85–91.
2. Rabenau, H.F. *et al.* (2005) Stability and inactivation of SARS coronavirus. *Med. Microbiol. Immunol.*, **194**, 1–6.
3. Sagripanti, J.L., Carrera, M., Insalaco, J., Ziemski, M., Rogers, J., and Zandomeni, R. (2007) Virulent spores of Bacillus anthracis and other Bacillus species deposited on solid surfaces have similar sensitivity to chemical decontaminates. *J. Appl. Microbiol.*, **102**, 11.
4. Murray, P.F., Baron, E.J., Pfaller, M.A., Tenover, F.C., and Yolken, R.H. (eds) (1995) *Manual of Clinical Microbiology*, 6th edn, American Society of Microbiology, Washington, DC.
5. Murray, P.F., Baron, E.J., Jorgensen, J.H., Pfaller, M.A., and Yolken, R.H. (eds) (2003) *Manual of Clinical Microbiology*, 8th edn, American Society of Microbiology, Washington, DC.
6. Flemming, D.O. and Hunt, D.L. (eds) (2006) *Biological Safety, Principles and Practices*, 4th edn, American Society of Microbiology, Washington, DC.
7. Ratula, W.A., Cole, E.C., Thomann, C.A., and Weber, D.J. (1998) Stability and bactericidal activity of chlorine solutions. *Infect. Control Hosp. Epidemiol.*, **19**, 323.
8. Rutala, W.A. and Weber, D.J. (1997) Uses of inorganic hypochlorite (bleach) in health-care facilities. *Clin. Microb. Rev.*, **10**, 597.
9. Hassan, K.A., Skurray, R.A., and Brown, M.H. (2007) Active export proteins mediating drug resistance in Staphylococci. *J. Mol. Microbiol. Biotechnol.*, **12**, 180.
10. Vali, L., Davies, S.E., Lai, L.L., Dave, J., and Amyes, S.G. (2008) Frequency of biocide resistance genes, antibiotic resistance genes and the effect of chlorhexidine exposure on clinical methicillin-resistant Staphylococcus aureus. *J. Atimicrob. Chemother.*, **61**, 525.

8
Fumigation of Spaces

Nigel Silman

This chapter deals with the use of gaseous disinfectants used to fumigate spaces, such as laboratory, animal housing, and primary biological containment systems. Firstly, we start with some definitions of what is meant by fumigation of spaces, why it is undertaken, and what is the result of fumigation.

8.1
Definitions

Fumigation does not:

- render clean
- sterilize
- kill everything
- remove vegetative cells
- remove spores
- get into inaccessible areas.

It does reduce the number of cells in a space, that is, it provides a "knockdown" in the number of viable organisms, rendering an area safe for untrained personnel to enter. The degree of knockdown achieved varies, but as a rule of thumb, a knockdown of 10^6 is generally deemed to be acceptable. Fumigation should be performed when commissioning and validating a new laboratory facility, before any work being started to ensure that in the event that emergency fumigation is required, the room is sealable. Fumigations are also routinely performed when servicing is required, following a spill of infectious agent and when required to knockdown background contamination, for example, the presence of fungal spores in a tissue culture laboratory. Finally, fumigation is always performed in laboratories where biological agents have been handled before decommissioning such a facility.

Working in Biosafety Level 3 and 4 Laboratories: A Practical Introduction, First Edition.
Edited by Manfred Weidmann, Nigel Silman, Patrick Butaye, and Mandy Elschner.

8.2 Practicalities

In order to carry out a fumigation of any space, the first requirement is that the space must be sealable, that is, there must be a containment envelope that will prevent loss of the gaseous fumigant. The *containment envelope* is generally defined as the line around a room or suite of rooms where containment measures apply, and it is an absolute requirement for the facility to be sealable at this line.

There are a number of fumigant chemicals in routine use, some with better efficacies than others. Probably the most widely used fumigant is formaldehyde vapor, although other frequently used chemicals include vapor-phase hydrogen peroxide (VHP), ethylene oxide, chlorine dioxide (Figure 8.1), and peracetic acid. Within Europe, fumigants used have to comply with EU emissions regulations that currently allow release of all fumigants into the atmosphere, although this situation is likely to change within the next few years. Some (e.g., VHP) result in the production of oxygen and water and thus do not release toxic chemicals into the environment, making their use attractive.

As formaldehyde is the most commonly used fumigant, this will be used as an example hereon (Figure 8.2). Formaldehyde is used by mixing with water (the ratio is dependent on the intended use) and boiling off the liquid in a "kettle." The resulting formaldehyde vapor is a surface decontaminant and works by fixing proteins. The effectiveness of formaldehyde as a fumigant is dependent on the time that it is in contact with the surface, the room humidity and temperature, the volume of space, and the presence of absorbant materials. A calculation is made before a first fumigation, which dictates the volume of fumigant required based on the room volume. The amount of formaldehyde used is calculated as 100 ml formalin (stock solutions are 38–40% formaldehyde in water) plus 900 ml water per 28 m^3.

Formaldehyde vapor does not penetrate surfaces, it does not reach inaccessible places, and it is nonreactive with the majority of laboratory materials; however, it is readily absorbed by materials such as paper and cardboard, which should be minimized in laboratories that are fumigated regularly. Fumigation failures can occur as a result of a number of factors; commonly these are leakage of the fumigant, insufficient hold/contact time, too cold, ventilation left on, too much absorbent material present, or the pots fail to boil off.

8.3 Fumigation Process

Before commencement of the actual fumigation process, a number of things have to be done within the space to be fumigated. Firstly, the less clutter that is present, the more effective will be the fumigation, as clutter means increased surface area and hence a greater volume of fumigant is technically required. Therefore, time to declutter! Remove everything that can be safely removed by surface disinfection,

Figure 8.1 A typical chlorine dioxide generator. (Activ Ox from Feedwater, http://www.feedwater.co.uk/, with kind permission from Feedwater).

including all waste tins to the autoclave, gowns (again remove to the autoclave before laundering) and finally remove all chlorine-containing materials. There is a remote possibility that chlorine-containing materials can react with formaldehyde to form the carcinogen *bis*-chloromethyl ether. If this should occur, then it would appear as a dense white "cloud," thus, it is best to avoid this possibility by the removal of the chlorine-containing materials – we usually place them in a sealable carry tin that can be surface disinfected to remove it from the laboratory.

Depending on the purpose of the fumigation, you may want to defrost fridges and freezers as well as open up drawers and cupboards. This would be done when, for example, the laboratory was being decommissioned or taken out of routine use for an extended period of time. When fumigating before a routine service interval, this sort of rigor is not needed and cupboards, drawers, freezers, and so on may be sealed to prevent accidental opening by service engineers. Pots need to be suitably placed in the space to be fumigated; it is a good idea to divide the fumigant volume

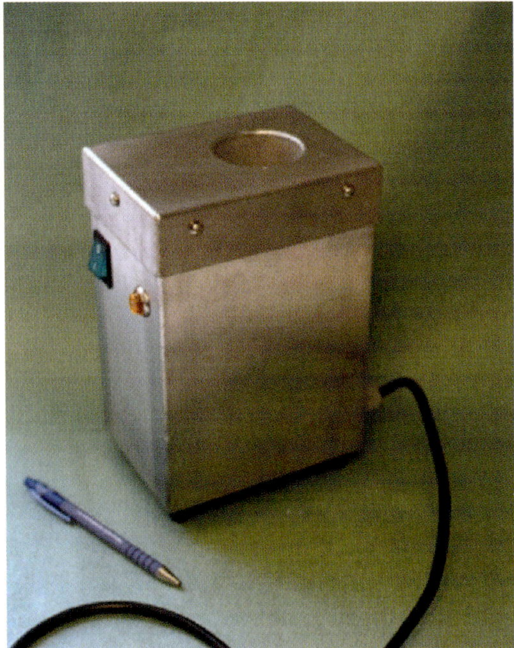

Figure 8.2 A typical "kettle" used to vaporize formaldehyde.

into two and use two separate pots to boil it off – this improves permeation of the gaseous fumigant as well as providing some resilience should one pot fail, for example. At this stage, spore strips should be placed strategically around the space. Place these in the more accessible areas, on the basis that if a $>10^6$ knockdown is achieved in hard to reach areas, then the fumigation will have been fully successful. Many laboratories now have "remote" fumigation systems, which generally means that there are special electric sockets (usually of a different color to identify them), which can be switched on and off from outside the laboratory envelope. Even if your laboratory does not have these, it is not a great problem, as the relatively large volume of formaldehyde will take a considerable time to come up to the boil (think how long your kettle takes to boil a few cupfuls of water). Switch on the pots and finally seal the door. You did remember to switch off the smoke alarms, didn't you? The room ventilation should now also be switched off and you should return to the area in about 1 h to monitor for formaldehyde leakage using a Formaldemeter. Before you get to this stage, you also need to complete the fumigation documentation and place warning notices on the laboratory door as well as at any entrances to the corridor where the laboratory is situated. Warning notices should detail the laboratory name/number being fumigated with the time and date started as well as the responsible person and their contact telephone number. In this way, should someone enter the area while the fumigation is underway, they will be aware of the event, and if there are any problems (e.g., leakage), they will know who should be contacted.

8.4 Validation of Fumigation

In order to determine whether fumigation has been successful, biological indicators are commonly used (Figure 8.3). These normally comprise strips impregnated with spores of *Bacillus atrophaeus* var. *globigii* at a concentration of 10^8 spores. Fumigation is a knockdown of at least 10^6, so these spore strips give an indication of this level of knockdown and ideally all spores are killed indicating a 10^8 or better knockdown. Spore strips are incubated for 48 h, after which a pass/fail result may be issued. There is anecdotal evidence that longer incubation of spore strips may give rise to growth of spores at a low level, possibly indicating spore damage rather than an absolute kill. A good first indication of a successful fumigation is that the pots are empty the next morning when the spore strips are removed from the area! Fumigation with formaldehyde vapor also results in condensation of the aqueous

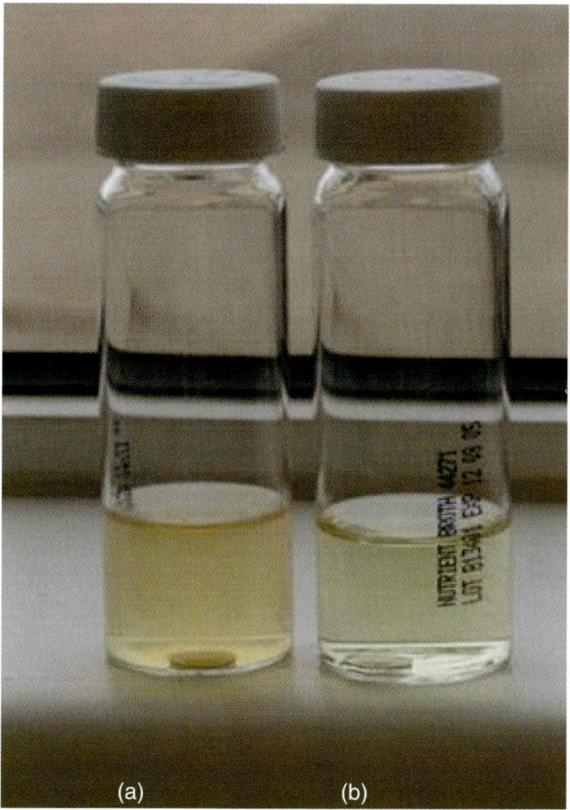

Figure 8.3 Biological indicators in growth medium. (a) Growth is shown in the culture and (b) fumigation pass.

liquid on all surfaces; thus, another indicator is the presence of moisture on walls and windows.

8.5 Post-Fumigation

Once the pots have finished boiling off to produce formaldehyde vapor, the space being fumigated should be exposed to the vapor for at least 12 h; this is called the *hold* or *contact time*. Step one after overnight fumigation (it does not have to be overnight, but this is a convenient way of ensuring a hold time of at least 12 h with minimum disruption to other areas and staff) is to reinstate the ventilation system. Some laboratories have a specific "vent" setting that runs the heating ventilation and air-conditioning (HVAC) system at a higher air change rate to clear the residual fumigant more quickly. In any event, it will take between 24 and 48 h to completely clear the fumigant and render the area safe to enter without respiratory protection; this time period fits in nicely with the spore strip incubation period of 48 h. To retrieve the spore strips during the immediate period after fumigation, workers will have to use personal respiratory protective equipment (RPE) and should always enter these areas in pairs. It is recommended that disposable Tyvek™ or similar disposable clothing is used to enter the area and workers should be trained in the correct use of respirators, as the levels of formaldehyde will be extremely high in the fumigated area. Bear in mind that respiratory protection should have chemical filters for this application.

The area may now be entered and spore strips recovered. Pots should be checked for complete vaporization of the fumigant; if the pots are not empty, it is very straightforward to repeat the fumigation while everything is setup correctly. There is no need to do anything else at this stage, as the room will take up to 48 h for the level of formaldehyde to decrease to below 1 ppm, the workplace exposure limit (WEL). Monitor the area around the door once the spore strips have been retrieved, to determine the level of formaldehyde escape into the corridor, which has occurred as a result of entering the laboratory; hopefully, this should be low as the HVAC system is drawing air into the area.

Once you have the spore strip results, and I hope that yours show a knockdown of at least 10^6, then you can sign off the fumigation paperwork as a "pass" and engineers and other service personnel may be allowed into the area to perform their service activities. One last thing to attend to before the area is serviced is to remove the pots, clean up any condensate on the floor and walls and give the area a thorough wash with water. This will remove *para*-formaldehyde residue that forms as a white powder during fumigation. Beware although, washing down can cause further leaching of formaldehyde, so be prepared to use RPE to undertake this task and monitor levels frequently. Another point to be aware of is that formaldehyde takes longer to be removed from underbench areas than from the rest of the room, so beware of residual pockets of fumigant. Oh, and do not forget to reinstate that smoke detector now, when you raise all of the paperwork to allow the service personnel into the laboratory.

8.6 Fumigation of Cabinets

Fumigating any contaminated space is essentially the same. The major difference with cabinet fumigation (Figure 8.4) is that generally a higher concentration of formaldehyde and a shorter hold time are used for the sake of convenience. For cabinets and other "small" areas, a ratio of one part formaldehyde to one part water is used with a hold time of only 6 h. Using this modified protocol, cabinets may easily be turned around overnight and be ready for use again first thing in the morning. Generally, spore strip indicators are not used for cabinet fumigations because the fumigations are validated initially using spore strip indicators; three successful fumigations will ensure that subsequent fumigations do not need to be monitored.

Figure 8.4 Fumigation setup of a biological safety cabinet class I.

8.7 Emergency Plans

Before actually starting a fumigation for the first time, you need to work out and (and write down) what you are going to do in the event that it all goes horribly wrong. Now, the main issue with fumigations that do not go according to plan is that there is a leakage somewhere and the fumigation needs to be aborted. The emergency plan needs to deal with this likelihood as well as other situations, for example, what you will do if the pots fail, you forget to inactivate the fire alarm (formaldehyde and other fumigants will cause the fire alarm to trip). In the first case, if when you monitor you find that formaldehyde is leaking extensively, then you need to be able to abort the fumigation. Step one is to turn off the remote fumigation sockets if you have them to stop further vaporization of fumigant; if you do not have remote sockets, then there is little alternative but to let them boil off. Step two is to reinstate the HVAC system, as that will then start to draw air into the area and prevent escape of fumigant as most of it will be drawn out through the extract high-efficiency particulate air (HEPA) filters. All you can then do is let time do its job as the fumigant levels will start to fall. You may need to isolate the corridor area outside of the laboratory being fumigated to prevent other staff entering the danger zone. In the event of failures of pots, there is no alternative but to repeat the fumigation, but use different pots this time.

8.8 Conclusions

Fumigation is not rocket science. An experienced and successful fumigator knows that preparation of the area by following a well-designed and tested checklist results in problem-free fumigations. Follow all of the steps logically and do not forget that you are never on your own; if you are not sure, ASK. Remember to notify someone (safety department, occupational health, line manager, etc.) if it all goes wrong, many heads are better than one to solve problems!

Summary

Fumigation of spaces using gaseous disinfectants is a necessary and required element of a BSL-3 laboratory. This chapter describes the theory and practicalities of undertaking fumigation of contaminated spaces be it a laboratory, biosafety cabinet, or piece of equipment.

9
Learning from a History of Laboratory Accidents

Manfred Weidmann

9.1
Introduction

A recent literature review on laboratory-acquired infections (LAIs) concluded that deviation from general "good microbiological practice" is the most frequent cause for LAI, and that training for compliance to procedures and regulations appears to be the best method to avoid these [1].

The major accident categories (no ranking) are technical failures (filters corrupted, centrifuge gasket leaky), failure of personal protection (protection of eyes, skin, and inhalation), and unsafe procedures (wrong strains, benchwork, spills, and sharps).

How safe is it then working in a biosafety laboratory? The answer is it is as safe as you make it to be. Anybody working with highly pathogenic organisms wants to walk home healthy at the end of each day. It is therefore in your own interest and of course that of your colleagues, your institution, and your community that you constantly think about the way you handle the organisms you are dealing with.

In the following sections, we follow the workflow in a laboratory from receipt of a strain via laboratory procedures to animal experiments. At each step, examples of published accidents are used to sharpen the readers' perception for the points that need consideration.

9.2
Strains

Do you really know what infectious organism and which strain you are working with? Is the information or are the data you received about the strain reliable, do you have any means to confirm it? Especially in recent years, several accidents came to notice where misconceptions about the strain being used in experiments played a role. In 2004, three researchers fell ill with tularemia in Boston, USA, after being exposed to wild-type *Francisella tularensis*, which was mixed up with the

Working in Biosafety Level 3 and 4 Laboratories: A Practical Introduction, First Edition.
Edited by Manfred Weidmann, Nigel Silman, Patrick Butaye, and Mandy Elschner.

vaccine strain the researchers thought they were working with. The mixture had inadvertently been supplied by another laboratory [2].

Nobody fell ill when a live *Bacillus anthracis* strain instead of an inactivated strain was delivered to a research group in Oakland, USA. This example is particularly striking because actually the strain was directly injected into mice and they all died within 3 days. This experiment was repeated and only after the entire second batch of animals died, *B. anthracis* was isolated from the dead mice [3].

The third case to be mentioned here is the case of a PhD student who came down with a mild form of severe acute respiratory syndrome (SARS) (as in 80% of all SARS infections) working with a West Nile virus (WNV) culture that in retrospect appeared to have been contaminated with severe acute respiratory syndrome coronavirus (SARS-CoV) through equipment that had not been properly decontaminated. The WNV work started 6 days after previous SARS-CoV work [4]. The potentially most widespread distribution of an unwanted virus strain occurred in 2009 when an experimental replication-deficient influenza A (H3N2) batch contaminated with an H5N1 strain was sent out to laboratories in several countries. In animal experiments, this batch rapidly killed ferrets, which led to the discovery of the contamination [5].

These examples push the point that any strain and presumed vaccine strains should always be verified and that decontamination procedures should be in place and appropriate (Chapter 7). It is a good idea to request a confirmatory test result on the identity of the strain from the shipper of the new strain, which on receipt you might still want to confirm in your own laboratory.

9.3 Eye Protection

These days it is good practice to wear eye protection, because it is known that, for example, pus squirting out of a lymph gland of a tuberculous guinea pig might hit your eye [6]. The elastic strap of a sterile mask might snap your eye as it happened to a technician checking mice 7 days after intracerebral inoculation with *Neisseria gonorrhoeae*. A severe eye infection developed subsequently [7]. A yolk-inoculum mix may squirt into your eye from a needle blown off a syringe while infecting yolk sacs of chick embryos with *Chlamydia trachomatis* [8].

Apart from the obvious, all these examples also point out that one must be careful with one's movements. What do I touch when? Do I adjust my mask strap with a glove that has just been inside the BSC (biosafety cabinet) handling an infectious agent? Do I force the plunger of the syringe although I feel a resistance indicating a blockage? What kind of syringe am I using anyway? A Luer-Lock syringe not available at the time of the report might have avoided the yolk squirt.

From my experience, I think that concentration and a state of raised awareness during your work in a BSL-3/4 (biosafety level) laboratory is of prominent importance and the best safeguard. Actually and you may smirk at this, it gives work in the BSL-3 laboratory a meditative character that gives a positive boost to your work but at the same time it makes this type of work exhausting after a couple of hours. Therefore, be alert to your level of awareness and limit the time spent in one shot to the extent you can handle.

9.4
Necropsies, Animal Experiments, and Sharps

It goes without saying that necropsies carry an inherent accident risk, as sharps have to be used. In 2002, two accidents during necropsies were reported from the Unites States leading to mild cases of West Nile fever. In one case, a thumb was lacerated with a scalpel while removing the brain from an infected blue jay; in another case, a needle stick occurred when harvesting the brain of an infected mouse. Both accidents occurred while working in a class II cabinet. In both cases, working in BSL-3 conditions would not have made a difference and in fact in order to detect WNV infections in a timely manner in diagnostic settings, work in BSL-2 conditions had been permitted [9].

The infamous mouse brain passage is very often the only means of isolation for arboviruses. Special precautions to reduce puncture wounds from syringes that could be taken are the use of finger cots or needle puncture-resistant gloves made from modern Kevlar-like fibers (e.g., Turtleskin [10], Amorflex [11]) (Figure 9.1). Another possibility is using syringes with spring driven retractable sheaths that fully cover the needle, slide back during the injection, and recoil into the protective position after drawing out the needle from the puncture site. Larger auto-resheathing syringes of this type (e.g., Sterimatic [12]) have been used for mass vaccination campaigns by veterinarians since 1982; however, in our experience, they are not

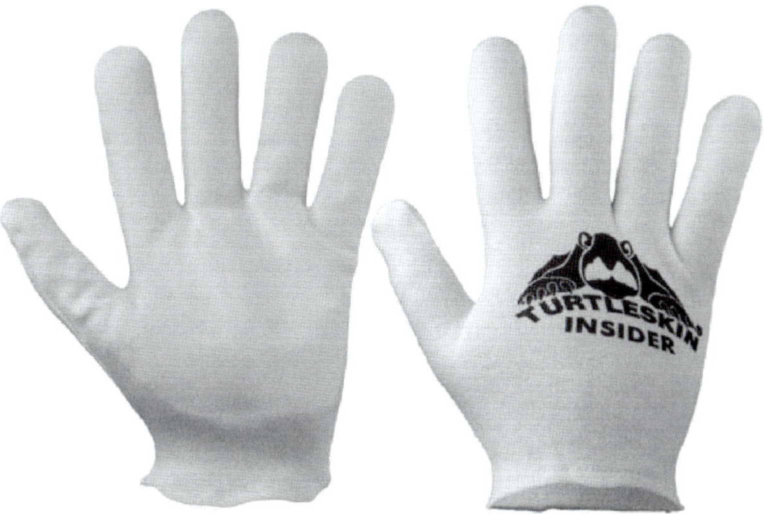

Figure 9.1 Turtleskin insider safety glove liners.

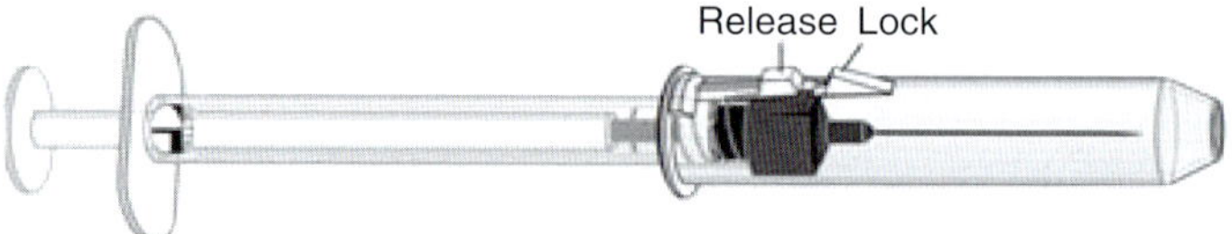

Figure 9.2 Example of a syringe with retractable sheath that fully covers the needle.

helpful when inoculating small mammals. An alternative are smaller 5–15 ml syringes using a technically different but in effect similar mechanism (e.g., Gettig Guard [13]) (Figure 9.2). The optimal combination for the project on hand has to be discussed and decided on in your team.

9.5 Skin Protection

Section 9.4 automatically brings us to the topic of protecting your skin and general behavior in a BSL-3 environment. Almost everybody will have experienced how during the first sortie into a BSL-3 laboratory all dressed up in gowns, gloves, and goggles, the incredible itch will jump you and challenge your resolution. Most people learn to deal with this after a while. Unfortunately, however, lapses can occur, as shown by cutaneous anthrax, which developed in the nape of the neck of a laboratory worker from a private laboratory that had been commissioned to deal with part of the overwhelming amounts of environmental samples that had to be analyzed after the anthrax attacks in the United States in 2001. The victim of the enormous workload was laboratory discipline [14].

One of the general rules of laboratory discipline of course is to cover up small wounds and not to work if a skin disease is troubling you. The latter was ignored by a technician with severe dermatitis on hands and forearms, who performed serology on material from SIV (simian immunodeficiency virus)-infected monkeys without gloves and on SIV-infected cultures with gloves in a class II cabinet. He seroconverted without disease as shown by HIV-2 positivity in a Western blot [15].

In another incident, covering a shaving wound and wearing gloves while subsequently handling vials with fresh *B. anthracis* isolates would have been a good idea to avoid cutaneous anthrax [16].

Covering a cut wound on the finger inflicted by a coverslip might have helped to avoid a local lesion of vaccinia developing into a mild generalized vaccinia infection. It is safe to assume that in this case inoculation of the cut occurred much later during continued work with vaccinia virus in the laboratory as finger cut and infection that initially surfaced as a pimple on the injured finger were 12 days apart, whereas the interval between inoculation and effects from inoculation of the vaccinia vaccine is known to be only about 3–4 days.

Another striking turn of this case was that quote: "she squeezed the pimple and a drop of pus squirted onto her face" more exactly onto her chin where a second lesion developed [17] (Figure 9.3). These accounts make clear that covering up

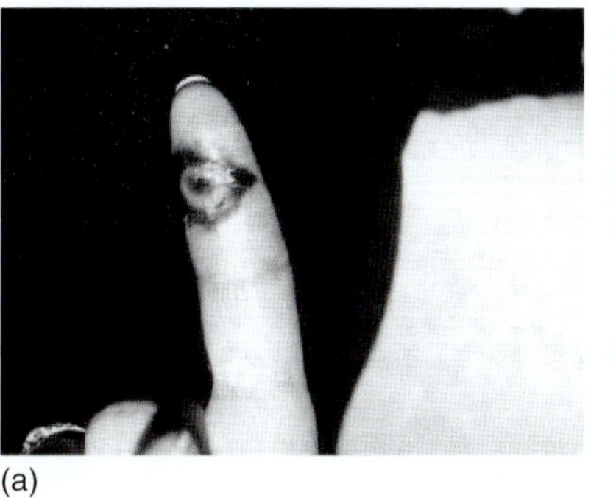

(a)

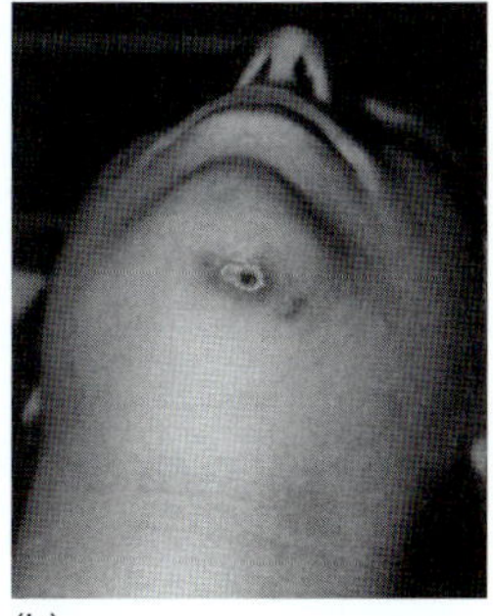

(b)

Figure 9.3 Skin lesions on finger and chin caused by Vaccinia virus. (Reprinted from [17] with permission from Elsevier.)

minor wounds should be taken seriously and that it is good laboratory etiquette to point that out to your colleagues if they are being negligent. In addition, you should be more watchful about minor cuts that might occur and the consequent inflammation that may result.

9.6 The Omnipresence of Aerosol

The following LAI shows clearly why even organisms that are not considered as serious threats may pose a problem when handled on the open bench. In 1997, an experienced medical laboratory scientific officer suffering from a sore throat and a cold infected himself when handling a heavy suspension of a toxigenic strain of *Corynebacterium diphtheriae* on the open bench, which he had received for an external quality assessment of biochemical tests. He developed a severe tonsillitis and was treated with antibiotics [18].

Working with new and as yet unidentified agents needs great care. In 1994, material from a seriously ill and dying patient in São Paulo was used to isolate the new Sabia virus via mouse brain passage. To properly characterize the new isolate, material was sent to Belem (3000 km away) to perform serology grouping of the potential new agent. Here, a technician was infected fell ill and survived [19].

Even agents that have not yet been proved to be transmissible by the aerosol route need attention as the history of smallpox shows. The transmission of smallpox by aerosol had been suspected for about 100 years until proof of aerosol transmission emerged in hospital outbreaks in Simmerath and Meschede in Germany in 1962 and 1970. Here, index cases caused nosocomial transmission chains in spite of being isolated. Especially in Meschede, retrospective smoke experiments showed that a staircase and a kitchen elevator acted like chimneys disseminating the virus. Remarkably driven by convection currents from radiators below the windows, a thin layer of smoke was observed to flow out of the partially opened window along the outer surface of the building until it entered the opened windows of upper floor wards. Altogether, the smoke distribution pattern was identical to the case distribution pattern in the hospital [20].

A laboratory accident gave the final proof when some scientists handled smallpox on the open bench at the department of medical microbiology of the Birmingham Medical School in the United Kingdom in 1978 one year after the last official smallpox case had been registered by the WHO. A laboratory photographer working one floor above the research laboratory became infected, fell ill, and died [21]. Air currents through a service duct connecting both floors were found to be responsible for the transmission [22].

Earlier work of Wedum [23, 24] had clearly shown that aerosols are easily generated by routine laboratory practices. He collected air next to typical actions in a laboratory and determined the colony count from the collected samples (Table 9.1). He showed, for example, that inserting a hot loop into a culture flask, and inserting a bacteria covered flammable loop into a flame actually disperses

Table 9.1 Aerosols in the laboratory.

Procedure	Colonies obtained per operation
Removing tight cover of standard Waring blender immediately after mixing culture	(1)
Opening lyophilized culture tube	86
Decanting centrifuge fluid into flask	17
Inserting hot loop in culture flask	9
Removing dry cotton plug from shaken culture flask	5
Pipetting 1 ml of inoculum to poured agar Petri plate	3
Pipetting 1 ml of culture into 50 ml of broth	1
Accident	**Colonies obtained per accident**
One 50 ml tube breaking in centrifuge and culture splashing side of centrifuge; air sampled 7 in. above centrifuge	1183
One 50 ml tube breaking in centrifuge but all 30 ml of culture staying in trunnion cup	4
Accidental breaking one ampoule of lyophilized nutrient broth culture on floor; air sampled at nostril height 18 in. each side of accident site for 1 h	491
Drop of culture falling 12 in. onto steel surface; air sampled within 2 ft of site	16
Petri plate cultures dropped on floor; air sampled 4 ft above floor, 70 ft from accident	9

(Tables reproduced from [25].)

bacteria. Using a sterile, single-use plastic loop therefore appears to be a good choice to avoid that.

All this work eventually led to the introduction of BSCs in the microbiological laboratory (Chapter 3). These high-efficiency particulate air (HEPA) filter equipped working spaces of course need maintenance and it should be regularly checked whether filters are still performing according to the manufacturer's recommendations [26]. In most countries, there are no regulated exchange periods. In the case of blocked filters however the laminar flow air intake of the front grill can be disrupted and actually blow in the face of the bench worker – not nice to imagine.

9.7 Centrifugation

Centrifuges in a BSL-3 laboratory should have gaskets to seal the centrifuge interior off from the outside. Ideal are centrifuges with removable cups fitted with transparent lids and gaskets that can be removed from the centrifuge and opened in a BSC. However, these gaskets need to be maintained as leaky gaskets can be the source of a finely disseminated, highly concentrated aerosol cloud of hantavirus [27], or a milligram of botulinum toxin [28]. In the first case, laboratory workers fell ill; however, in the second case, they did not. The latter accident is a reminder that toxin contamination is not the same as intoxication!

More recently, a centrifuge accident at a Texas laboratory involved genetically engineered H3N2 and H5N1 influenza virus strains. The accident happened in a BSL-3 laboratory that was run as a "BSL-3 plus" laboratory with workers additionally wearing positive pressure respirators or even suits. Although this helped in the case of the centrifuge accident, the question may be asked how this potent infectious virus got categorized into "BSL-3 plus" in the first place [29].

9.8 Spills

Carrying around things in the laboratory as, for example, the aforementioned removable centrifuge cups brings us to the topic of laboratory spills.

Let us start from the rear end. What happens in the case of a spill? Do you have a routine in place in your laboratory on how to deal with a spill? Bang!, here it is, something horrible spilled all over the bench or the floor. The first impulse is to reach out for some disinfectant and decontaminate the visible spill. It is all too human, to want to clean and to quickly cover up a spill, but this is actually reckless behavior toward your colleagues and can lead to further even worse consequences.

What about the aerosols then and would not it be wiser to let the system do what it is designed for. In a normal BSL-3, there will be at least about 10 room air exchanges an hour and that through HEPA filters! In many laboratories therefore apart from immediate personal decontamination and removal of gloves and gowns on the way out, people are drilled to leave the site of immediate aerosol danger and after caring for themselves call for help. Until a team has arrived dressed up for the occasion, the BSL-3 air filter system will have done its work and surface decontamination can proceed at an easy pace. Have you talked this through in your team and do you have an emergency plan?

Glassware of course is not used in BSL-3 laboratories anymore and that is why you cannot cut yourself on broken culture tubes of a rack that fell to the floor to infect yourself with *Mycobacterium tuberculosis* [30]. On the other hand, spills do still occur.

In 2003, a spill occurred in a detachable waste lock of a BSL-3 cabinet that could not be reached with the attached gloves from the inside. Eventually, the lock was sprayed from the inside with alcohol (inappropriate for SARS-CoV) and opened from the outside after 10 min, which led to an infection of the laboratory worker, who subsequently recovered [31].

The samples of the Sabia virus mentioned earlier were also sent to the Arbovirus Research Unit at the Yale School of Medicine, where a year later a virologist fell ill after cleaning up a spill in a centrifuge in a BSL-3 laboratory. A tissue culture supernatant contained in a 250 ml bottle had cracked. He cleaned it up wearing a gown, gloves, and just a surgical mask. He did not report the incident, which came to light when he fell seriously ill but fortunately too survived [32]. An unreported spill with *B. anthracis* and presumably an inefficient spill cleanup inside a BSL-3 laboratory at Fort Detrick in 2002 led to the subsequent spread of the spores to at least the locker room adjacent to the BSL-3 laboratory [33].

The last two examples of instinctive cover up behavior is the incarnation of BSL biosafety culture going very wrong. The safety of the team working in the BSL-3 laboratory depends on a set of agreed rules and a culture of open communication and trust. The rules are there exactly because safe practices are not instinctive and therefore need to be agreed on adapted to ongoing projects and adhered to by everybody in the team.

9.9 Laboratory Accident Statistics

Looking at the statistics of several efforts trying to record LAI, the following picture evolves. In surveys, from 1924 to 1977 and 1980 to 1991, 4454 LAI with 175 deaths were recorded [1]. Harding and Byers [34] covered 1979–2004 and recorded 1148 LAI with 36 deaths. Pike [35] found that 59% of the LAI occurred in research and 17% in diagnostic laboratories and Harding recorded 50.8% of accidents in research and 44% in diagnostic laboratories. In Pike's survey, the cause of 82% of the accidents was unknown. Being in and around the laboratory as a source for LAI decisively points toward infectious aerosols. As described earlier, work done by Wedum inspired by Pike's first analyses [36–38] clearly pointed toward the dangers from aerosols. This also explains why most accidents occur with highly pathogenic organisms or those that need a very low infectious dose [35, 39, 40].

Only 18% of LAI were classified as accidents with an overall mortality rate of 4%. However, among the classified accidents, needlestick accidents lead, together with accidents involving spills and spray at a range of about 25% each followed by sharps (14%), aspiration through pipettes (13%, not common anymore), and bites or scratches by animals or ectoparasites (14%).

Looking at the agents mainly involved in LAI, a shift can be observed from the top six of *Brucella* sp., *Coxiella burnetti*, hepatitis viruses, *Salmonella* sp., *F. tularensis*, and *M. tuberculosis* in Pike's survey covering 1930–1978 [39] to *M. tuberculosis*, arboviruses, *C. burnetti*, hantaviruses, *Brucella* sp., and hepatitis viruses, in a survey covering 1979–2004 [34]. These changes also reflect the changes of incidence of these diseases. Looking at clinical microbiological diagnostics alone (2002–2004, United States), the top hits are *Shigella* sp., *Brucella* sp., and *Salmonella* sp. [41].

In summary, it can be said that the most common exposures occur by inhalation of infectious aerosols generated by accidents or working procedures, percutaneous inoculation, and contact of contaminated material with mucous membranes. An accumulative demonstration of all of these points was recently performed by a veterinary student in South Africa who suffered from West Nile fever after a necropsy where he used a bone saw while removing the brain (WNV infects horses and is neurogenic) from a pony just wearing gloves. He must have inhaled aerosols from the bone saw action, suffered no percutaneous injury but presumably contaminated the mucous membranes in his mouth or in the eye [42].

Summary

Laboratory accidents are rare and documented accidents are even rarer. This chapter tries to summarize a history of published accidents to discuss the biosafety and microbiological practice issues the accidents raised. The reader is taken through the steps from sample or stock arrival through the various steps of processing and analysis and tries to discuss safety issues by using accidents examples.

References

1. Kimman, T.G., Smit, E., and Klein, M.R. (2008) Evidence-based biosafety: a review of the principles and effectiveness of microbiological containment measures. *Clin. Microbiol. Rev.*, **21**, 403–425.
2. (2004) ProMED-mail. TULAREMIA, LABORATORY-ACQUIRED - USA (MASSACHUSETTS) 2002.
3. CDC (2005) Inadvertent laboratory exposure to Bacillus anthracis–California, 2004. *MMWR Morb. Mortal. Wkly. Rep.*, **54**, 301–304.
4. Senior, K. (2003) Recent Singapore SARS case a laboratory accident. *Lancet Infect. Dis.*, **3**, 679.
5. (2002) ProMED-mail. AVIAN INFLUENZA, ACCIDENTAL DISTRIBUTION - CZECH REPUBLIC ex AUSTRIA (03), 2009.
6. Tegström, A. (1942) A case of accidental laboratory tuberculous eye infection. *Acta Tuberc. Scand.*, **16**, 16330–16333.
7. Diena, B.B., Wallace, R., Ashton, F.E., Johnson, W., and Patenaude, B. (1976)

Gonococcal conjunctivitis–accidental infection. *Can. Med. Assoc. J.*, **115**, 609–635.

8. Smith, C. (1958) Accidental laboratory infection with trachoma. *Br. J. Ophtamol.*, **42**, 721–722.
9. CDC (2002) Laboratory-acquired West Nile virus infections–United States, 2002. *MMWR Morb. Mortal Wkly. Rep.*, **51**, 1133–1135.
10. Turtelskin Gloves *https://www.turtleskin.com/* (accessed 13 April 2013)
11. Hexamor Gloves *http://www.hexarmor.com/* (accessed 13 April 2013).
12. Sterimatic Safety Needle System *http://www.sterimatic.com/* (accessed 13 April 2013).
13. Gettig Guard Safety Needle System *http://216.92.52.175/guard.html* (accessed 13 April 2013).
14. (2002) ProMED-mail. ANTHRAX, HUMAN, LABORATORY WORKER - USA (TEXAS).
15. Khabbaz, R.F., Rowe, T., Murphey-Corb, M. *et al.* (1992) Simian immunodeficiency virus needlestick accident in a laboratory worker. *Lancet*, **340**, 271–273.
16. CDC (2002) Suspected cutaneous anthrax in a laboratory worker–Texas, 2002. *MMWR Morb. Mortal Wkly. Rep.*, **51**, 279–281.
17. Wlodaver, C.G., Palumbo, G.J., and Waner, J.L. (2004) Laboratory-acquired vaccinia infection. *J. Clin. Virol.*, **29**, 167–170.
18. (1998) ProMED-mail. Diphtheria, laboratory accident - UK.
19. Lisieux, T., Coimbra, M., Nassar, E.S. *et al.* (1994) New arenavirus isolated in Brazil. *Lancet*, **343**, 391–392.
20. Wehrle, P.F., Posch, J., Richter, K.H., and Henderso, D. (1970) Airborne outbreak of smallpox in a German hospital and its significance with respect to other recent outbreaks in Europe. *Bull. World Health Organiz.*, **43**, 669–679.
21. 1978) Smallpox in Birmingham. *Br. Med. J.*, **2**, 837.
22. Shooter, R. (1980) *Report of the Investigation into the Cause of the 1978, Birmingham Smallpox Occurrence*, Her Majesty's Stationery Office, London.
23. Wedum, A.G. (1961) Control of laboratory airborne infection. *Bacteriol. Rev.*, **25**, 210–216.
24. Wedum, A.G. (1964) II. airborne infection in the laboratory. *Am. J. Public Health Nations Health*, **54**, 1669–1673.
25. Wedum, A.G. (1964) Laboratory safety in research with infectious aerosols. *Public Health Rep.*, **79**, 619–633.
26. U.S. Department of Health and Human Services, Centers for Disease Control and Prevention, and National Institutes of Health (2007) *Biosafety in Microbiological and Biomedical Laboratories*, 5th edn, U.S. Government Printing Office, Washington, DC.
27. Centers for Disease Control and Prevention (1994) Laboratory management of agents associated with hantavirus pulmonary syndrome: interim biosafety guidelines. *MMWR Recommendations Rep.*, **43**, 1–7.
28. (2006) ProMED-mail. Botulinum toxin, laboratory exposure - USA (MA).
29. The Sunshine Project The Bird Flu Lab Accident that Officially Didn't Happen (2007) *http://www.sunshine-project.org/* (accessed 11 January 2007).
30. Gale, G.L. (1957) Accidental infection with tubercle bacilli in laboratory technicians. *Can. Med. Assoc. J.*, **76**, 646–648.
31. Normile, D. (2004) Infectious diseases. Second lab accident fuels fears about SARS. *Science*, **303**, 26.
32. CDC (1994) Arenavirus infection–Connecticut, 1994. *MMWR Morb. Mortal Wkly. Rep.*, **43**, 635–636.
33. (2002) ProMED-mail. Anthrax, laboratory Exposure - USA (Md).
34. Harding, A.L. and Byers, K.B. (2006) *Biological Safety: Principles and Practices*, 4th edn, ASM Press, Washington, DC.
35. Pike, R.M. (1976) Laboratory-associated infections: summary and analysis of 3921 cases. *Health Lab Sci.*, **13**, 105–114.
36. Sulkin, S.E. and Pike, R.M. (1949) Viral infections contracted in the laboratory. *N. Engl. J. Med.*, **241**, 205–213.
37. Sulkin, S.E. and Pike, R.M. (1951) Survey of laboratory-acquired infections. *Am. J. Public Health Nations Health*, **41**, 769–781.

38. Sulkin, S.E. and Pike, R.M. (1951) Laboratory infections. *Science*, **114**, 3.
39. Pike, R.M. (1979) Laboratory-associated infections: incidence, fatalities, causes, and prevention. *Annu. Rev. Microbiol.*, **33**, 41–66.
40. Sewell, D.L. (1995) Laboratory-associated infections and biosafety. *Clin. Microbiol. Rev.*, **8**, 389–405.
41. Baron, E.J. and Miller, J.M. (2008) Bacterial and fungal infections among diagnostic laboratory workers: evaluating the risks. *Diagn. Microbiol. Infect. Dis.*, **60**, 241–246.
42. (2009) ProMED-mail. West Nile Virus - South Africa: 2009, From Necropsy.

10 Bridging the Gap between Requirements of Biocontainment and Diagnostics

Manfred Weidmann, Frank T. Hufert, and Nigel Silman

As discussed in Chapter 6, specimen transport is subject to international regulations that try to assure that specimens are transported safely, timely, and efficiently. Once the specimen has arrived at the diagnostic laboratory, biosafety has to be maintained to protect the staff that receive and unpack the specimen. Experience shows however that although the IATA (International Air Transport Association) regulations exist, specimens sent to specialized laboratories, especially from less organized countries, do tend to arrive in all but the expected safe state. When unpacking samples, one has therefore to expect that vials may have been damaged or absorbent materials to safeguard for such a situation is of the wrong type, too little, or nonexistent. Anecdotal incidents of sharps (unsheathed syringe needles sticking in cotton wool) being sent in, or stories about ingenious packing devices made of wood and nails are well known in the BSL-3/4 (biosafety level) diagnostic laboratory scene.

Generally, all samples sent in for routine diagnostics are considered as infectious with an unknown etiological agent. Therefore, good microbiological practices and procedures that help to protect the staff receiving and unpacking the samples are required. Additional structural biocontainment (engineering controls, e.g., use of a microbiological safety cabinet) can be used but concepts of unpacking vary. Unpacking in a biosafety cabinet (BSC) class I in a BSL-2 laboratory is standard in the United Kingdom while unpacking in a BSC class II in a BSL-2 laboratory is routine in Germany even in routine diagnostic laboratories.

Most laboratories will unpack all samples in a BSL-2 laboratory and some will unpack in a BSL-3 laboratory using a BSC class II or even a BSC class III cabinet if a particular agent is suspected. At this point, however, it may be pointed out again that a BSL-3 laboratory itself does not provide an added protection to staff. It is a structural biocontainment to protect the environment from infectious agents getting out of the laboratory. Even in a BSL-3 laboratory, the best safeguard is good microbiological practice, that is, safe procedures supplemented with basic personal protective equipment (PPE) (gowns, gloves, and goggles) either with or without the use of additional engineering controls such as biological safety cabinets.

As a rule, it may be said that clinical samples contain lower concentrations of agent, are of smaller volume, and are therefore potentially less hazardous than, for

Working in Biosafety Level 3 and 4 Laboratories: A Practical Introduction, First Edition.
Edited by Manfred Weidmann, Nigel Silman, Patrick Butaye, and Mandy Elschner.

example, bacterial or viral cultures and their supernatants, which contain much higher titers of infectious agents. The operator protection factor (OPF) provided by BSC class I, II, and III cabinets are 10^5, 10^5, and 10^6, respectively. These protection factors will easily suffice for the concentrations of most infectious agents in clinical samples, allowing for the fact that it is very unlikely that these types of samples will be readily aerosolized.

A good indication of how safe microbiological procedures are when dealing with a highly infectious BSL-4 agent is the outbreak of Crimean–Congo hemorrhagic fever virus, which recently erupted in Turkey and resulted in more than 4000 cases since 2002. The only accidents that occurred were due to a needlestick obtained when taking a blood sample from a patient and when sampling blood without wearing gloves. Accidents did not happen in any of the diagnostic BSL-1 and BSL-2 laboratories that dealt with the outbreak samples [1, 2].

Nucleic acid diagnostics are very often the best choice for samples of unknown origin coming in directly from acute cases, as long as the samples turn out to be from the acute viremic/bacteremic phase (otherwise these assays are likely to yield a negative result). In recent years, several publications have shown that chaotropic guanidine-isothiocyanate-containing buffers in the inactivation buffers of commercial nucleic acid extraction kits do effectively inactivate most agents [3].

For serological analysis, which is most useful when the diagnostic window has moved on to where nucleic acid detection is useless, most laboratories inactivate sera at 56 °C for 30 or even 45 min and this seems to be a safe way of inactivation. Attention should however be drawn to a publication that demonstrated that, especially for some hemorrhagic fever viruses, 60 °C for 60 min is more effective [4].

As to the strategy to be used to keep procedures and manipulations to a minimum for unclear or suspicious samples, it is a good idea to perform nucleic acid detection assays first, wait for the results, and then undertake the serological assays. Serology of course could be performed immediately in a BSL-3 laboratory in the case of urgency; however, using the appropriate inactivation procedures, a BSL-2 laboratory will allow safe handling of these samples with sufficient worker protection.

In summary, a risk assessment must be made when routinely handling clinical samples for diagnostic testing. There needs to be a thorough appraisal of the sample types and unpacking procedures when samples are received. Most likely, this will allow unpacking in a BSL-2 laboratory, except for BSL-4 virus-containing samples (known or presumed). Samples suspected of presenting an aerosol risk to workers should be unpacked in a BSC class I or II (or even class III for some samples). If there is any indication while unpacking samples that there is damage or leakage, then these should be transferred immediately to a BSC to give adequate protection for the staff members.

Summary

Which considerations are necessary to handle clinical samples of cases with fever of unknown origin with a potentially containing BSL3/4 agents? Various approaches are discussed.

References

1. Gurbuz, Y., Sencan, I., Ozturk, B., and Tutuncu, E. (2009) A case of nosocomial transmission of Crimean-Congo hemorrhagic fever from patient to patient. *Int. J. Infect. Dis.*, **13**, e105–e107.
2. Tutuncu, E.E., Gurbuz, Y., Ozturk, B., Kuscu, F., and Sencan, I. (2009) Crimean Congo haemorrhagic fever, precautions and ribavirin prophylaxis: a case report. *Scand. J. Infect. Dis.*, **41**, 378–380.
3. Blow, J.A., Dohm, D.J., Negley, D.L., and Mores, C.N. (2004) Virus inactivation by nucleic acid extraction reagents. *J. Virol. Methods*, **119**, 195–198.
4. Mitchell, S.W. and McCormick, J.B. (1984) Physicochemical inactivation of Lassa, Ebola, and Marburg viruses and effect on clinical laboratory analyses. *J. Clin. Microbiol.*, **20**, 486–489.

11
Risk Assessment Procedures

Åsa S. Björndal

11.1
Introduction

A risk assessment procedure is the process of evaluating the biological risk(s) arising from a biohazard, for example, handling a strain of Ebola virus or *E. coli*, considering the adequacy of any existing controls (e.g., availability of a protective equipment and laboratory containment), and deciding whether or not the risk(s) is acceptable (based on the availability of protective measures).

Every laboratory environment needs a solid routine for assessing and managing risks posed by handling pathogens and other hazardous materials. A risk assessment procedure ensures a safe work environment for the laboratory personnel (including laboratory staff, contractors, and maintenance and service personnel), for individuals in the close surroundings (including personnel sharing the facility and visitors) as well as the environment. It also ensures a robust research environment without cross-contaminations and other problems corrupting research results.

Laboratory incidents/accidents involving biological hazardous material could result in human disease. Potential other consequences could be risks for the environment (establishment/displacement/interactions of new organisms) and also lead to economic loss and decreased credibility and reputation of the institute/organization.

A biological risk assessment aims at identifying not only biosafety risks but also biosecurity risks. Inadequate biosafety management could result in unintentional exposure or laboratory-associated infection of personnel or accidental release to community/environment. Laboratory biosecurity risks could result in malicious use of biological agent or misuse of critical relevant information such as research results or information of patient data.

Working in Biosafety Level 3 and 4 Laboratories: A Practical Introduction, First Edition.
Edited by Manfred Weidmann, Nigel Silman, Patrick Butaye, and Mandy Elschner.

11.2
Risk Identification

To be able to assess all potential risks, knowledge is needed including detailed facts about the intrinsic properties of the infectious agents, information about work procedures and methods to be used, information about who will handle the material (students, senior laboratory staff, etc.), and finally knowledge on how all these factors interact. In addition to biological risk, other risks such as chemical, mechanical, electrical, and radiation risks/hazards may have to be considered.

11.2.1
Timing of Assessment

The risk assessment should be performed before laboratory procedures are carried out, for instance, before starting a new project or activity or before introducing significant changes, for example, when new hazardous biological materials are introduced, ahead of introducing novel work procedures/routines and at renovation and moving of laboratories. It is also useful to map and evaluate potential risks involved before introducing new pieces of equipment, such as cryostats, flow cytometric machines, aerosol chambers, and other apparatus that needs special consideration because of the ability of creating aerosols or increasing risk for blood-borne transmission.

Risk assessment should be continuously improved in a revision process. Importantly, revision of risk assessment strategies should also be performed after the occurrence of a laboratory incident/accident. A laboratory incident investigation will then identify potential failing of the original plans for managing biological safety and security issues and indicate how to improve future risk management.

11.2.2
A Qualitative Risk Assessment

Biological risks are difficult to quantify and therefore the procedure and subsequent result, based on all available processed information, will be more qualitative in nature. A qualitative assessment of risk should include all handling activities including storage, transfer/transport/shipment, and destruction of the hazardous material. To some extent visualize the biological risk of handling a given biological agent, a risk matrix (Figure 11.1) may be of use. Depending on the nature of the material (e.g., infectious, attenuated, or inactivated) and the handling procedure, the positioning in the matrix may vary. For instance, culturing of live *Mycobacterium tuberculosis* in a BSL-3 laboratory may pose a "very high risk," whereas performing a polymerase chain reaction (PCR) analysis of inactivated genetic material from the same bacterium may be of "medium risk" or even "low risk."

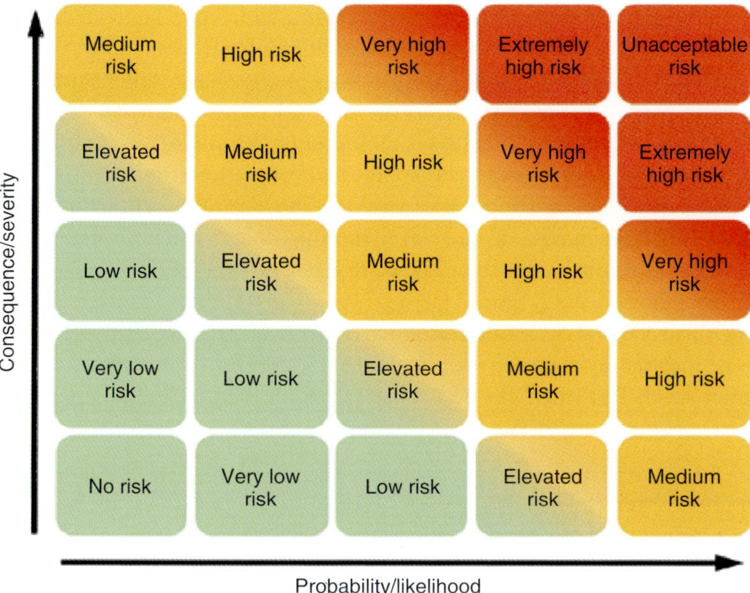

Figure 11.1 Risk matrix with "acceptability levels."

11.2.3 Systematic Documentation

As the risk assessment procedure should be incorporated already when planning the laboratory activity, some laboratories have prepared special risk assessment templates to be filled in and further processed. The risk assessment procedures should involve the following three general steps:

1) *Identification of all potential hazards/risks* – The existence of risk has to be perceived and identified. A wide range of questions could be asked. A few examples are given below:
   (i) Which pathogens? Transmission route (natural/laboratory)? Infectious dose? Virulence? Pathogenicity? Are there known laboratory-associated infections?
   (ii) Are there other hazards to consider, for example, the use of hazardous chemicals, inflammables, isotopes, or laboratory animals?
   (iii) What volumes and concentrations of infectious material will be used?
   (iv) Which methods and laboratory procedures will be used?
   (v) Is there a risk for aerosol generation?
   (vi) May the use of needles and sharps be limited?
   (vii) Does all laboratory staff have enough knowledge and are they well trained, for example, competent? Are there certain risk groups to consider?

(viii) Are prophylactic measures available (e.g., vaccination) and health surveillance system in place? Can a potential infection be treated?

(ix) Are there validated methods for inactivation of infectious material?

2) *Analysis and evaluation of risk(s) involved* – Processing of all available information. Can the risk of direct or indirect contact, risk of percutaneous exposure, and risk for exposure by inhalation or ingestion be further reduced? Laboratory work with biological agents will always involve some level of risk and it is very difficult to protect against every conceivable event. In addition, resources for risk mitigation are never infinite. A decision on whether the risk is acceptable or not has to be taken. The risk matrix could here be of use in the decision process (Figure 11.1).
3) *Management of risk* – Decide on additional protective measures that are needed to prevent or reduce the risk, for example, laboratory containment levels (i. e. biosafety level) BSL1 to BSL4 and protective equipment (including personal protective equipment/PPE).

The risk assessment procedure is a teamwork and should be performed by the laboratory worker (researcher, research student, laboratory technician, etc.) together with project leader or other senior liable individual.

The form is to draft an introductory paragraph on the nature of the experiments or changes in procedures planned. The paragraph should include information on biological material as well as handling procedures. Then draft a list of questions (e.g., similar to the list (a)–(i) earlier) and document the answers in a structured way. After having done this for a few times, you may come up with your institutional biosafety questionnaire that you then can fill out in a more simple and convenient way (Figure 11.2). If needed, additional expertise could be provided by consultations of biosafety specialist, engineering staff, general safety specialist, and so on.

Risk assessments should be documented, continually revised to fulfill actual needs, and stored in a way so that they are easily accessible to all personnel that need them. Some institutes/organizations require that documented risk assessments are evaluated in a biosafety committee.

11.3 Additional Points for General Risk

In a general biosafety risk assessment, influencing factors of individual work tasks, workload, work pace, and work hours should be considered. In addition, organizational issues such as social climate, cooperation, leadership, and the possibility to influence as well as ergonomic issues (e.g., illumination and access to proper equipment) will affect the outcome of the risk management strategy and should therefore be given an evaluation beforehand.

Risk assessment strategies tend to focus more on biological risks managed by general biosafety programs in place. The laboratory biosecurity risk assessment, on the other hand, may be an integrated part of the overall risk assessment procedures and thus consider structural and physical securities, for example, facility design

Risk Assessment form

Department, unit:
Project number:
Title and description of the project:
Date:

Biological agent (known or suspected):				
Detailed description (if applicable):				
Risk Group (agent)	2 ■	3 ■	4 ■	
Lab Biosafety Level	2 ■	3 ■	4 ■	Comments:
Licence (agent)	No ■	Yes ■	If yes, valid until (Date):	

GMM* included?	No ■	Yes ■	* genetically modified microorganisms
Licence (GMM)?	No ■	Yes ■	If yes, valid until (Date):

				Comments
1	Known or suspected human pathogen?	Yes ■	No ■	
2	Are the chosen strains virulent?	Yes ■	No ■	
3	May non-pathogenic or less virulent strains be used?	Yes ■	No ■	
4	Route of transmission: natural?			
5	Route of transmission: in the laboratory?			
6	Infectious dose? (i.e. the relative amount of micro-organisms required for infection)	High ■	Low ■	
7	Laboratory associated infections (LAI) reported or documented? If yes, state reference	Yes ■	No ■	
8	Are there any prophylactic measures available for the staff? If yes, which?	Yes ■	No ■	
9	Are there diagnostic methods available at suspected LAI? If yes, provide details	Yes ■	No ■	
10	Risk of allergies? (e.g. animals) Particular risks for certain risk groups (e.g. pregnant women)?	Yes ■	No ■	
11	Are personnel informed of the seriousness of a potential LAI?	Yes ■	No ■	
12	Is there treatment* available at exposure/infection? If yes, how? *according to established medical practice	Yes ■	No ■	
13	Maximum volume (or similar) infectious material handled at one time (litre, cm^3)? Enter concentration (low/high)			
14	Any risk for splashes? If yes, type of splash protection	Yes ■	No ■	
15	Any risk for aerosol generation?	Yes ■	No ■	

Figure 11.2 Example of a risk assessment form.

16	Will centrifugation be used? If yes, identify risks. Centrifuge cups?	Yes ☐	No ☐	
17	Will needles/sharps be used?	Yes ☐	No ☐	
18	Will pressurized systems be used? (e.g., vacuum suction, gas cylinders)	Yes ☐	No ☐	
19	Will sonication be used? If yes, identify risks	Yes ☐	No ☐	
20	Other risks (e.g., heat, cold) or risk procedures (e.g., transport, animal handling)?	Yes ☐	No ☐	
21	May risk procedures be replaced by low risk procedures?	Yes ☐	No ☐	
22	Disinfectant(s) to be used?			
23	Are the rooms appropriate? Consider separation, vicinity to equipment, etc.	Yes ☐	No ☐	

24	Need of specific protective measures to be taken? Describe					
25	Do all personnel have appropriate training? Specify	Yes ☐	No ☐			
26	Are there risks to the technical/service personnel or others in the room?	Yes ☐	No ☐			
27	Work in biosafety cabinet (BSC) or isolator? If yes, specify	I ☐	II ☐	III ☐	Isolator ☐	No ☐
28	Which personal protective equipment should be used? (e.g., respiratory protection, gloves, face shield)					
29	Is there cause for enhanced security measures on bio-threat?	Yes ☐	No ☐	If yes, contact Biosafety Officer		

30	Work with *hazardous* or *inflammable* chemicals? If yes, which?	Yes ☐	No ☐	
31	Work with radioactive isotopes? If yes, which?	Yes ☐	No ☐	

32	Additional information:			
33	Project plan is attached (always for new risk assessments presented to the Biosafety Committee)	Yes ☐	No ☐	
34	Documents attached (state appendix number) ☐ Standard Operating Procedures (SOPs) ☐ Reports ☐ References LAI ☐ Other references ☐ Other documents			
33	Discussed in the Biosafety Committee	Yes ☐	No ☐	If yes, date:
34	Discussed with Health & Safety representative	Yes ☐	No ☐	If yes, date:

Persons participating in the Risk Assessment (Name, affiliation):		
Project leader:		Date and signature:
Planned revision of the Risk Assessment		Date:

Figure 11.2 (*Continued*)

with controlled laboratory access, protective measures, and the presence of a contingency plan. However, biosecurity risk and threat assessments may need a more multidisciplinary approach involving for instance law enforcement agencies. When performing a biosecurity risk assessment, assets, such as strains of pathogens and research data, are evaluated and potential threat(s) in the environment are identified. The biosecurity risk handling strategy aims at safeguarding valuable biological materials from unauthorized access.

Summary

To enable a safe and secure laboratory environment, procedures and practices to identify, analyze, and manage biological risks need to be in place. Handling biological materials in research, diagnostic, or industrial laboratories requires compliance to national and international (EU) regulations covering workers health protection as well as quality standards or guidelines for laboratory procedures and methodology. A risk assessment procedure should be established, implemented, and maintained in every laboratory setting. Laboratory management commits to the continuous provision of the supporting tools such as risk documentation and risk communication systems. If such system fails to be implemented, consequences can be great for human or animal health, the environment, and the liability of the institute or organization. The risk assessment procedure is a shared responsibility between the laboratory staff and the organization's management and should be an integrated part of planning of all laboratory activities providing guidance on how to handle identified hazards and mitigate biological risks.

In this chapter, we discussed various steps to be taken, from perception of risk through procedures for hazard identification to a systemically documented assessment of biological risk. An example template is also provided that may be used as an example on assessment and documentation of biological risk in a laboratory environment.

Further Readings

CEN Laboratory Biorisk Management, CWA 15793, *ftp://ftp.cenorm.be/CEN/Sectors/TCandWorkshops/Workshops/CWA15793_September2011.pdf* (accessed 13 April 2013).

CEN (2012) Laboratory Biorisk Management – Guidelines for the Implementation of CWA 15793, CWA 16393:2012 (published in January 2012), *ftp://ftp.cen.eu/CEN/Sectors/List/ICT/Workshops/CWA%2016393.pdf* (accessed 13 April 2013).

WHO (2004) *Laboratory Biosafety Manual*, World Health Organization, Geneva, *http://www.ebsaweb.eu/ebsa_media/Downloads/Resources/Biosafety7.pdf* (accessed 13 April 2013).

WHO (2006) *Biorisk Management: Laboratory Biosecurity Guidance*, World Health Organization, *http://www.ebsaweb.eu/ebsa_media/Downloads/Resources/WHO_CDS_EPR_2006_6.pdf* (accessed 13 April 2013).

12
Biosecurity

Jürgen Mertsching

12.1
Introduction

The term *biosecurity* has been used with multiple meanings. In the animal industry, the term *biosecurity* refers to the protection of an animal colony from microbial contamination. In some countries, it is used in place of the term *biosafety*. The difficulty of finding distinct definitions for biosafety and biosecurity begins with the fact that these English terms cannot be adequately translated into other European languages. The two terms would be translated, for example, as *Biosicherheit* in German or *Biosécurité* in French. For the purpose of this chapter, the definitions based on the WHO and CWA Laboratory Biorisk Management guidelines [1–3] are used to describe biosafety and biosecurity as separate concepts with a number of common aspects:

- *Laboratory biosafety* describes the containment principles, technologies, and practices that are implemented to prevent the unintentional exposure to biological agents and toxins or their accidental release.
- *Laboratory biosecurity* describes the protection, control of, and accountability for biological agents and toxins within laboratories, in order to prevent their unauthorized access, loss, theft, misuse, diversion, or intentional unauthorized release.

Biosafety and biosecurity components are often integrated into a biorisk management program. The term *biorisk*, which embraces both concepts, describes the combination of the probability that damage will occur and the potential impact of this damage by a biological agent. To what extent biosecurity measures are needed depends on the critical biological materials and the kind of work conducted with those materials. At least for the handling of biological material of risk groups 3 and 4 and of organisms and toxins, which are specified in national lists [4–6], biosecurity procedures should be in place. While the implementation of biosecurity measures should be more stringent in institutions handling large amounts of critical biological material, for example,

Working in Biosafety Level 3 and 4 Laboratories: A Practical Introduction, First Edition.
Edited by Manfred Weidmann, Nigel Silman, Patrick Butaye, and Mandy Elschner.

reference centers for the respective high threat pathogens, their implementation in research laboratories either handling small amounts of critical biological material or isolated genes or gene products may not be necessary at the same level.

In addition, biosecurity applies not only to dangerous pathogens and toxins but also to biological material that may not be a potential threat but nonetheless have a certain value for the organization [2].

12.2 Biosecurity as Part of a Biorisk Management Program

A biorisk management program enables an organization to effectively identify, monitor, and control laboratory biosafety and biosecurity aspects of its activities. To be effective, it needs to be conducted using a structured systematic approach, integrated throughout the organization [3].

Even though the objectives of biosafety and biosecurity are not the same, there are joint measures to achieve these goals. Both are based on risk (threat) assessment, access controls, employee qualification, and responsibility. Both involve inventory lists, controlling the use of biological agents, documentation of agent transfer and shipment, and emergency concepts [7]. Biosafety measures are often equally effective in the area of biosecurity. In addition, there are specific biosecurity aspects, which are not related to employee or environmental protection.

The following key components regarding biosecurity have to balance safety, security, and accountability of high threat pathogens and toxins while sustaining day-to-day scientific research.

12.3 Risk (Threat) Assessment Process

Risk assessment is a systematic, structured process for analyzing and determining risk. In the field of biosafety and biosecurity, risk is known as a *function of probability* and *consequences* [8]. Here, the term *risk* can be defined by answering the questions "What can go wrong?," "How likely is it?," and "What are the consequences?"

For biosecurity, the risk assessment varies with:

- the likelihood of theft of a biological agent;
- the severity of the consequences of an attack with that agent or the consequences of the loss of valuable material or information.

For carrying out an assessment of misuse risk, different models exist. The following structure contains a four-step process [9].

12.3.1 Identify and Prioritize Biological Materials

In a first step, the biological materials existing at the institution have to be identified. This includes all forms of the material, including nonreplicating materials (e.g., toxins), their location, and quantities.

An examination of the potential for misuse of these biological materials and of the consequences of their misuse has to follow. The biological materials have to be prioritized based on the consequences of misuse (i.e., risk of malicious use).

12.3.2 Identify and Prioritize the Threat to Security of Biological Materials

An "insider threat" is seen as most likely to occur. Therefore, the types of "insiders" who may pose a threat to security of biological materials at the institution have to be identified. This knowledge may then be complemented by identification of the types of "outsider" who may pose a threat. The motives, means, and opportunity available to these various potential adversaries have to be examined.

12.3.3 Analyze the Risk of Specific Security Scenarios

The drawing up of a list of possible biosecurity scenarios or undesired events that could occur at the institution (each scenario is a combination of an agent, an adversary, and an action) is a prerequisite for the implementation of adequate threat mitigation measures. One should consider all possible means of access to the agent within the laboratory, how an undesired event could occur, the putting in place of protective measures to prevent an occurrence, and how the existing protective measures could be breached (i.e., vulnerable points).

The different scenarios are then evaluated as to the likelihood of their occurring and the ensuing consequences determined. Although a wide range of threats is possible, certain threats are more probable than others and not all agents/assets are equally attractive to an adversary. Consequently, the scenarios should be ranked and prioritized for action to be taken by management.

12.3.4 Integrate the Biosecurity Risk Assessment Process into a Biorisk Management Program

To be effective, the threat assessment process has to be integrated into an overall risk management program. It is the task and responsibility of the management board to develop a biosecurity plan to establish how the institution will mitigate against unacceptable risk. Management has to ensure that the necessary resources are available to achieve the protection measures derived from the threat assessment. Management must set a framework for training, daily implementation, and reevaluation of the biosecurity program on a regular basis.

12.4 Physical Security and Access Control

Physical security measures are the most obvious way to protect valuable biological materials from theft or misuse. These will, however, only be effective in combination with the other components of a biosecurity program.

Physical security is realized as graded protection. A fence may define the initial boundaries of the university campus or the area of a company. It surrounds property protection areas and is sufficient to protect low risk assets, public access offices, and open buildings. Certain buildings may be declared for limited access and may include laboratories with moderate risk assets. Hallways surrounding exclusion areas may form part of a second protection layer. High-risk assets are centralized in exclusion areas that may contain high containment laboratories as well as computer network centers. The more people are allowed to enter the graded security layers and move toward higher risk assets, the stricter the requirements for access control become.

The type of access control selected depends on the level of protection. Access controls may include (Figure 12.1):

- physical key;
- electronic key (e.g., card reader);
- positive identification by a guard;
- personal identification number (PIN); and
- biometric devices (e.g., finger print reader).

Access rights must be reexamined periodically and retracted if no longer required. When employment is terminated or a project in a high containment laboratory is finished, a proper procedure should be in place for revoking access rights, including the return of ID badges, keys, and cards.

In addition to locks, access control mechanisms often include intrusion detection devices, which alert security personnel in case of attempts to gain access without authorization. Alarms must be checked whether they are valid or false alarms. When alarms turn out to be correct, security personnel must have been trained to respond appropriately. A clear and fast response is an overall important element of biosecurity.

Video control measures have to be balanced against the principles of protection of data privacy and the statements of the workers union.

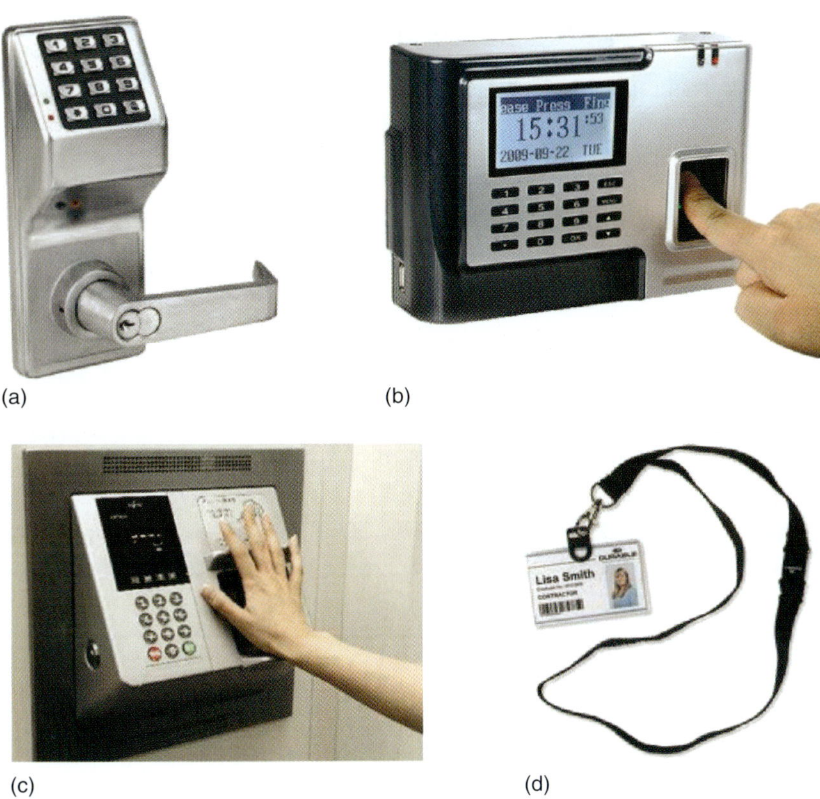

(a) (b) (c) (d)

Figure 12.1 Physical security and access control. (a) Electronic locks (Trilogy® DL4100 Electronic Digital Privacy Lock), (b) Biometrics Scanner (*http://omgtoptens.com/wp-content/uploads/2013/02/Biometric_Fingerprint_Time_Attendance_Scanner___Door_Access_System_-_OW69-01__84177_zoom.jp*), (c) (*http://www.netcharles.com/orwell/pics/fujitsu-palm-biometrics.jpg*), and (d) ID badges.

Regarding access control, the need for entry by visitors, laboratory workers, management officials, students, cleaning staff, and maintenance staff must be borne in mind.

12.4.1 Physical Security – Biosecurity Meets Biosafety

From a biosafety perspective, access control to laboratory space is effective in protecting personnel by limiting the number of individuals who may be exposed to a hazard. Access controls should be planned and implemented in awareness of the laboratory operations and biosafety practices. The installation of biometric readers within high containment areas may counteract biosafety practices. Fingerprint scanners, for example, on freezers, would require the personnel to remove their gloves when opening the freezer. Eye scanners may not work effectively where personnel have to wear eye protection. Biometric readers should be located at the entrance to the anteroom. Within high containment areas, the use of keys, magnetic cards, or PINs should be considered.

12.5 Material Management

Material management facilitates an environment that discourages individuals from stealing biological materials. The objective is to know what biological material is present in an institute, how and where the material is stored and handled, and who is responsible for it. Not all biological material has to be controlled in this way. A threat assessment procedure has to be undertaken for the organization to identify and prioritize biological material with threat potential or other material of high value to the organization.

Inventory control is key for accountable material management (Figure 12.2). Inventory control aims at having a comprehensive overview of all pathogens and toxins in an organization. The simple accumulation of pathogenic or toxic material with insufficient oversight and control creates a potential biosafety hazard for the employees and a security risk for the organization.

Filing biological materials is faced with the intrinsic problem that small amounts of viable material can reproduce to a larger quantity. Although inventories of pathogens and toxins may vary in complexity, they should include the following information [10]:

- types of material (name, strain, serotype, etc.);
- forms of material (solution or pellet, freeze dried, etc.);
- quantities of material (number of vials, amount of liquid, post-experiment quantities);

Figure 12.2 Material management. Inventory control (liquid nitrogen storage). (*http://www.google.de/imgres?imgurl=http://upload.wikimedia.org/wikipedia/commons/d/d6/Liquid_nitrogen_tank_for_cryostorage-06.jpg*).

- location of material (in short- or long-term storage or in use);
- contact or responsible employee;
- employees who have access to the material;
- modification of the original biological properties of material (e.g., genetically modified organisms);
- conformation, date, and method of destruction or inactivation of material; and
- dates of transfer of material (delivery and departure) and end-user or recipient receipts.

Although it is not practicable to determine the exact quantity of critical organisms at any given time, control of such material is greatly facilitated if it is confined to particular containers that can be tracked as separate items, for example, discrete, identifiable, and countable units [11]. Such items could be a container with test tubes or sealed ampoules containing biomaterial in solution or a freezer, provided that control measures are in place, which ensure item integrity. A competent person has to be appointed who is accountable for the material. Any anomalies observed by that person have to be reported to the responsible officials.

An information recording system must be established for biological materials. Depending on the kind of material, there could be a variety of information to be stored. Forms that are comprehensive and easy to complete should form the basis of the collection and administration of the information relevant for biosecurity. The resulting inventory should be regarded as sensitive information and kept in a secure and limited access database.

Good material management enables immediate detection of discrepancies, which is the basis for secure handling of critical material. It might help to exculpate an organization. In the event that a laboratory appears to be the source of a pathogen that has been used maliciously, inventory information can be used to defend the laboratory's practices.

12.5.1
Material Management – Biosecurity Meets Biosafety

It is good biosafety practice to document all organisms and different strains of bacteria and viruses in a structured manner. For genetic engineering work, it is of particular importance to list the genetically modified organisms so that the different stages of modification can be relocated in freezers and storage containers. Clear labeling of commonly used freezers and storage boxes for infectious material is necessary to avoid a mix-up of samples and to protect employees and colleagues. The biohazard sign for laboratories at biosafety level 2 (BSL-2) and higher is intended to provide notification of potential biohazards.

The goal of biosafety to inform all people around a given area about potential biohazards by pinning up a list of the organisms handled may conflict with biosecurity concerns. From a biosecurity perspective, such information must not be displayed openly. Inventory lists of BSL-3 and BSL-4 laboratories contain sensitive information and must be handled and stored securely by the responsible employee. Posting a biohazard sign with information about the employee to contact would meet both requirements.

12.6 Personnel Security Management

Security is fundamentally a people problem rather than a technological one. The anthrax mail attacks in 2001 were most likely executed by an insider [12]. Looking for measures to address the insider threat, personnel security management is clearly the most effective.

Mechanisms of personnel security management include:

- verifying credentials;
- checking references;
- requesting a criminal history from the authority; and
- conducting an in-depth background check.

Before a contract is handed out to a potential employee, credentials and past work experience and academic qualifications must be verified. For work in BSL-2 laboratories, checking references and recommendations may be sufficient. If the employee is going to be granted access to high-risk assets in exclusion areas of BSL-3 and BSL-4, more sophisticated background screening may be required. Such a background check should be performed together with a security specialist and on the basis of national law enforcement rules (e.g., the German Security Clearance Check Act, SÜG [13]).

Background checks can only reveal the history of an employee. In a current working situation, disputes between staff or personal problems outside the workplace have the potential of leading to the development of an insider threat. Although they may be difficult to detect, the laboratory manager should try early on to identify problems that could escalate into security problems.

Management is responsible to see that employees are qualified and able to work in the position they hold. Proactive monitoring of the state of mind and health of employees will reduce the number of biosafety and biosecurity-related incidents at a facility. Principal investigators and others who are responsible for operations in BSL-3 and BSL-4 laboratories should intervene when an individual does not appear to be in a suitable state for work. In exclusion areas of high containment laboratories, it is important that the person responsible is empowered to temporarily remove an individual from a work environment when that individual's health may impair safe and secure operations. This course of action should be documented in advance in the institution's safety policy and available for the employees so that the affected individual understands the basis for the action. The manager must be responsive to staff, and the institution should offer programs to help employees who may have questions or concerns about financial matters, mental health, or substance abuse.

12.6.1
Personnel Security Management – Biosecurity Meets Biosafety

By ensuring that members of the workforce are suitable for the position they hold, an organization can mitigate against the risk of both accidental and malicious acts. According to an institutional safety program, biosafety practices include the supervision of newcomers until the necessary skills and qualification are routinely demonstrated. Before a new employee is authorized to access areas of higher BSL, the supervisor has to be convinced that the work is performed safely and securely. Gradually, the work undertaken on the employee's own responsibility can be increased.

Working alone with hazardous material increases the risk of an incident or accident, especially at night or during the weekend. A strategy adopted both for biosafety reasons and to address the insider threat is to use a "two-person rule," for work in BSL-3 and BSL-4 areas.

12.7
Transport of Biological Materials

Organizations should have material transport policies in force governing the movement of valuable material within an institution and outside of a facility. Transport policies should address the need for appropriate documentation and control procedures for pathogens during transport. The organization must ensure that the procedures for the transport of critical biological material are established in accordance with legal requirements for the transport of dangerous goods [14, 15]. This applies in particular to material with the potential for dual-use. Under EU Council regulations, it is illegal for an EU member state to export any items listed in the regulation annex without authorization [16, 17]. The listed items include human, animal, and plant pathogens and toxins.

12.7.1
Transfer within an Institution

For internal transfer, dangerous pathogens and toxins are only allowed to be moved between restricted areas that provide the required BSLs. Such movement may occur as laboratories exchange materials under study, add materials to the inventory in case new materials are received, or send them to disposal areas (e.g., autoclave rooms).

Outside the restricted areas, biological material is more vulnerable to theft. Stringent measures and instructions have to be in place. Everyone who has access to dangerous biological agents in transport should be subject to the same personnel security requirements as those required for individuals with access to the material in the laboratory [18]. Instructions for the transport procedure should include the handling of the materials in areas that are used for temporary storage, such as shipping or receiving offices. Control measures should be implemented in these areas at a level equivalent to the restricted areas where the material is used and dependent on the threat assessment.

12.7.2 Transport Outside of the Facility

Transport security measures start with adequate communication between facilities before the material is dispatched on tour. Information on the transport schedule should be exchanged as well as telephone and contact data. The sending organization is responsible that only trained and competent personnel familiar with proper containment, packaging, labeling, and documentation handle the respective material.

The sender must ensure that the shipping company in charge has a valid license for handling the dangerous material. The company's policy on lost or stolen material must be known and how law enforcement representatives are contacted in such a case. Most commercial carriers provide a tracking system and a tracking number, which allows the package to be followed along the route during the external transport. Although this service is not in real time and does not guarantee custody over a package at all times, tracking provides information regarding the relative position of a package.

On delivery, the recipient should notify to the sender that the material ordered has arrived safely and completely.

12.7.3 Transport – Biosecurity Meets Biosafety

Exchange of biological materials is essential for diagnostic purposes and research in the life sciences. Shipping companies must provide efficient transportation of biological materials, especially frozen materials. Limiting the amount of time that the material is outside of a facility's restricted areas will also reduce the opportunity for theft.

Biosecurity measures aim to limit access to dangerous biological materials during transport and to meet biosafety requirements at the same time. If only a limited number of competent individuals are permitted to handle dangerous pathogens and toxins, the exposure risk will be reduced in the event of a spill.

12.8 Information Security

Information security aims to safeguard sensitive information that could be used by an adversary to enter restricted areas and to misappropriate pathogens, toxins, or other valuable materials. Organizations need to have a policy for identifying sensitive information and should have procedures in place for controlling access to such information. This includes information on [19]:

- pathogens and toxins;
- storage locations;
- entry codes;
- physical security; and
- security procedures.

The procedures for addressing information security should stipulate the secure storage of all sensitive written records and data, including electronic records, alongside thorough destruction of paper files to be discarded and complete erasure of unwanted electronic files.

It should be clarified with the IT department whether sensitive information should be stored on stand-alone computers or isolated networks within restricted areas to prevent unauthorized access to that information. There should be a strict directive regarding personal computers, laptops, storage media, cameras, and so on entering or leaving the facility. Sensitive information should not be carried on USB (Universal Serial Bus) memory sticks.

Depending on the threat potential of the biological material, the staff of the IT department with administrative access to the relevant computer networks should be incorporated into the process of background screening as an integral part of personnel security management.

12.9 Incident and Emergency Response Planning

Biosecurity policies should consider that unexpected situations, incidents, and emergencies can occur despite an effective biorisk management. Such eventualities should be dealt with proactively and contingency plans should set in place.

Access controls must be implemented in a manner that does not hinder emergency response. The procedures must ensure both emergency entry for responders and the security of the protected biological materials. Emergency plans should contain individually tailored procedures for minor incidents up to major accidents.

The biosecurity-driven response planning should be integrated with existing general emergency and contingency plans. It should address in addition [19]:

- damage to property;
- theft or loss of assets;
- inventory discrepancies;
- attempted or successful unauthorized access;
- violence and threats; and
- suspicious behavior.

Particular emphasis has to be paid to the identification of people to take charge during an emergency. The assignment of roles and responsibilities and the setting up of a chain of command require designation of authority to people with specific roles during an emergency. Internal personnel must be trained on how to react in such a situation, working in cooperation with external emergency services.

12.9.1 Emergency Response Planning – Biosecurity Meets Biosafety

Good biosafety practices aim to preserve the safety and health of laboratory employees and the surrounding community. Practical and effective protocols for emergencies must include plans for medical emergencies, facility malfunctions, fires, escape of animals from the laboratory, and other potential emergencies. Training in emergency response procedures must be provided for emergency response personnel and other responsible staff according to institutional policies.

In high containment areas with dangerous pathogens and toxins, emergency response measures should not allow an adversary to gain unauthorized access to critical biological materials by activating an alarm or overriding the access control. Documentation is necessary showing how the system is designed to meet both biosafety and biosecurity requirements in the event of an emergency.

Summary

Biosecurity is a fairly new term and is discussed as part of biorisk management. It includes risk assessment by identifying and prioritizing biological materials at a given facility and the analysis of specific security scenarios for them. Access control, material management, and personnel security management are features that need to be considered as well as the case of transportation of samples within or outside of a facility. Information security and incident and emergency response planning asks for merging aspects of biosecurity with biosafety.

References

1. World Health Organization (2004) *Laboratory Biosafety Manual*, 3rd edn, World Health Organization, Geneva.
2. World Health Organization (2006) *Biorisk Management: Laboratory Biosecurity Guidance*, World Health Organization, Geneva.
3. CEN (European Committee for Standardization) (2011) Laboratory Biorisk Management, CEN Workshop Agreement CEN CWA 15793, *ftp://ftp.cenorm.be/CEN/Sectors/TCandWorkshops/Workshops/CWA15793_September2011.pdf.*
4. CDC Select Agent Program, Centers for Disease Control and Prevention, US Department of Health and Human Services, Select Agent Program (2011) *http://www.cdc.gov/phpr/documents/DSAT_brochure_July2011.pdf* (accessed 13 April 2013).
5. The Australia Group Australia Security Sensitive Biological Agents List, *http://www.australiagroup.net/en/*

biological_agents.html (accessed 13 April 2013).

6. European Commission (2005) Technical Guidance on Generic Preparedness Planning – Interim Document, *http://ec.europa.eu/health/ph_threats/Bioterrorisme/keydo_bio_01_en.pdf* (accessed 13 April 2013).
7. National Institutes of Health/Centers for Disease Control and Prevention (2009) *Biosafety in Microbiological and Biomedical Laboratories (BMBL)*, 5th edn, CDC, Atlanta *http://www.cdc.gov/biosafety/publications/bmbl5/* (accessed 13 April 2013).
8. Salerno, R.M. and Gaudioso, J. (2007) *Laboratory Biosecurity Handbook*, CRC Press, Boca Raton, FL, p. 13.
9. BMBL, 5th, see [7], p. 107.
10. Clevestig, P. (2009) *Handbook of Applied Biosecurity for Life Science Laboratories*, Stockholm International Peace Research Institute (SIPRI), p. 8 *http://books.sipri.org/files/misc/SIPRI09HAB.pdf.*
11. World Health Organization (2006) *Biorisk Management: Laboratory Biosecurity Guidance*, World Health Organization, Geneva, p. 20.
12. United State Department of Justice (2010) Amerithrax Investigative Summary, *http://www.justice.gov/amerithrax/docs/amx-investigative-summary.pdf* (accessed 13 April 2013).
13. SÜG (1994) Law on Prerequisites and Procedures for Security Clearance Checks Undertaken by the Federal Government (Security Clearance Check Act). (Gesetz über die Voraussetzungen und das Verfahren von Sicherheitsüberprüfungen des Bundes (Sicherheitsüberprüfungsgesetzn) of 20 April 1994.
14. International Air Transport Association (IATA) (2012) *http://www.iata.org/ps/publications/dgr/pages/index.aspx* (accessed 13 April 2013).
15. UNECE (2011) European Agreement Concerning the International Carriage of Dangerous Goods by Road, (ADR).
16. EU (1994) Council Regulation (EC) No 3381/94 of 19 December 1994 setting up a Community regime for the control of exports of dual-use goods. *Off. J. Eur. Communities*, **L367** (31), 1.
17. EU (2009) Council Regulation (EC) No 428/2009 of 5 May 2009 setting up a Community regime for the control of exports, transfers, brokering and transit of dual-use items. *Off. J. Eur. Union*, **L134**, 29.
18. Salerno, R.M. and Gaudioso, J. (2007) *Laboratory Biosecurity Handbook*, CRC Press, Boca Raton, FL, p. 55.
19. Clevestig, P. (2009) *Handbook of Applied Biosecurity for Life Science Laboratories*, International Peace Research Institute (SIPRI), Stockholm, p. 20.

Appendix

Practical Course

Day 1

Dexterity – Training

Working in a BSL-3 (biosafety level) laboratory all dressed up and wearing three gloves requires a lot of preparation and concentration. To accustom you to the situation, a mock pipetting scheme mimicking a typical dilution series after a virus culture has to be done twice, once with colored solutions and once with clear solutions. This mock dilution series is pipetted as "blue-in-blue solution" in the first round and "clear-in-clear solution" in the second round.

1) Preparing supernatant stocks
   In a 50 ml tube, you will find 10 ml of mock virus culture supernatant. Pipette 5 × 1 ml into cryotubes and close the lids.
2) Dilution series
   To prepare the dilution series, add 900 µl of the "dilution buffer" into five 1.5 ml tubes.
   Start the dilution series by adding 100 µl of the "virus supernatant" to the first tube of the dilution series. Deposit the tip, take a new tip, mix the dilution gently by agitating with the pipette, and transfer 100 µl of the first dilution mixture to the next tube. Continue the dilution series down to tube five using a new tip for each step.

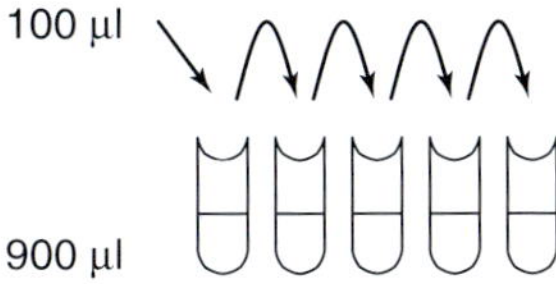

3) Inoculation of six-well plate
   Transfer 500 µl of the dilution buffer into the first six-well plate cavity (negative control). Transfer 500 µl of each dilution into a separate six-well plate cavity.

Working in Biosafety Level 3 and 4 Laboratories: A Practical Introduction, First Edition.
Edited by Manfred Weidmann, Nigel Silman, Patrick Butaye, and Mandy Elschner.

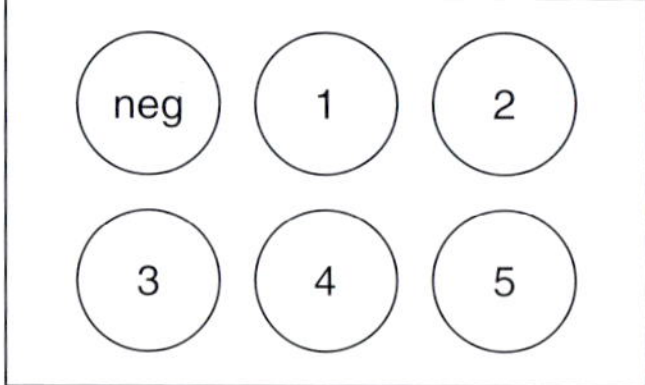

Repeat this protocol using the colorless reagents.

Material. Blue fluid: crystal violet solution and fluorescent solution (e.g., Dermalux).

Virus Inactivation

1) Virus culture inoculation
   You will be given a tube with 150 µl of VSV (vesicular stomatitis virus) culture with a titer of 1×10^{11} pfu ml^{-1}. Dilute this down to 1×10^{8} pfu ml^{-1} in three 1 : 10 dilution steps.
   To prepare the dilution series, add 900 µl of the dilution buffer [phosphate buffered saline (PBS)] into three 1.5 ml tubes. Start the dilution series by adding 100 µl of the virus supernatant to the first tube of the dilution series. Deposit the tip, take a new tip, mix the dilution gently by agitating with the pipette, and transfer 100 µl of the first dilution mixture to the next tube. Continue the dilution series down to tube three using a new tip for each step.

   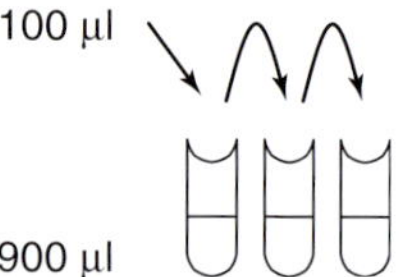

   You will also receive a small culture flask of 80% confluent VeroE6 cells. Remove the culture supernatant. Transfer 1 ml of the final VSV dilution into the flask using a 1 ml pipette and leave on cells for 30 min. Finally, add 9 ml culture medium.
   Place the culture flasks in an incubator at 37 °C.

Bacillus anthracis Culture

You will be given two tubes with a swab (swabs are frequently received from clinical laboratories or hospitals).

You will also receive two blood agar plates (Coloumbia Agar (CA)).

Transfer the material of each swab onto a plate. Apply the material on one site on the plate using the swab and after that use a loop to spread the material on the plate.

CA

Place the plates and tube in an incubator at 37 °C.

Material. Swab 1. *B. anthracis* A 58 (vaccine strains devoid of plasmids), *Bacillus cereus*.

Day 2

Dexterity – Training

Working in a BSL-3 laboratory all dressed up and wearing three gloves requires a lot of preparation and concentration. To accustom you to the situation, a mock pipetting scheme mimicking a typical dilution series after a virus culture has to be done twice once with colored solutions and once with clear solutions. This mock dilution series is pipetted as "blue-in-blue solution" in the first round and "clear-in-clear solution" in the second round.

1) Preparing supernatant stocks
   In a 50 ml tube, you will find 10 ml of mock virus culture supernatant. Pipette 5 × 1 ml into cryotubes and close the lids.
2) Dilution series
   To prepare the dilution series, add 900 μl of the "dilution buffer" into five 1.5 ml tubes.
   Start the dilution series by adding 100 μl of the "virus supernatant" to the first tube of the dilution series. Deposit the tip, take a new tip, mix the dilution gently by agitating with the pipette, and transfer 100 μl of the first dilution mixture to the next tube. Continue the dilution series down to tube five using a new tip for each step.

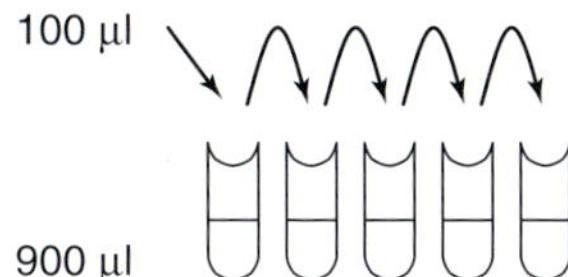

3) Inoculation of six-well plate
   Transfer 500 μl of the dilution buffer into the first six-well plate cavity (negative control). Transfer 500 μl of each dilution into a separate six-well plate cavity.

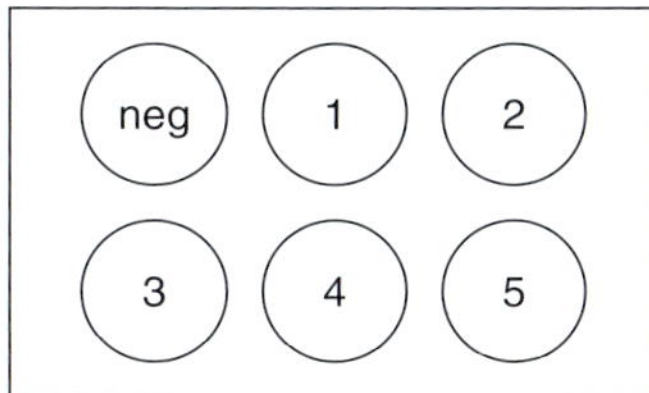

Repeat this protocol using the colorless reagents.

Virus Inactivation

To show how various disinfectants [Triton-X-100, sodium dodecyl sulfate (SDS), Tween-20] act on enveloped viruses, virus inactivation will be performed. In order to exclude toxic effects of the inactivating agent itself on the cell culture will be tested at a concentration of 1% in parallel to the inactivated virus supernatants.

The read out will be a $TCID_{50}$ assay, which indicates at what concentration of virus or in this case inactivated virus 50% of the cells are destroyed by virus replication.

1) Dilution of inactivating agents for toxic effect control
   a. You will find a tube with 10% solutions of the inactivation agents Triton-X-100, SDS and Tween-20 on your bench.
   b. Arrange three 1.5 ml tubes for each inactivation agent and mark them with the name of the inactivation agent and 1%. Add 450 μl culture medium to each tube. Add 50μl of Triton-X-100, SDS, Tween-20 10% to the respective tubes.

2) Inactivation of VSV culture supernatant
   a. Use a 10 ml pipette to retrieve the VSV culture supernatant from the flask and transfer it into a 50 ml tube. Centrifuge it at 3000 g/5 min to pellet the cell debris.
   b. Arrange three tubes for each virus inactivation dilution series (3 × 3 tubes) marking them with the respective dilutions and agent used (e.g., Triton 1%, 0.1%, and 0.01%). Add 450 μl "hot" VSV culture supernatant to tubes 1–3 of each dilution series.
   c. *For inactivation* with Triton-X-100, SDS, and Tween-20, start each inactivation dilution series by adding 50 μl of the inactivating agent (10%) to the first tube of the dilution series. Deposit the tip, take a new tip, mix the dilution gently by agitating with the pipette, and transfer 50 μl of the first dilution mixture to the next tube. Continue the dilution series down to tube three using a new tip for each step.

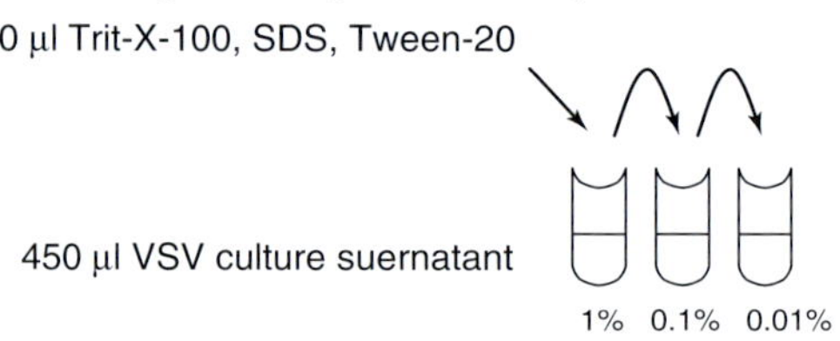

   d. *For the toxicity control,* simply add 450 μl cell medium (no virus!) to the 50 μl detergent remaining in each tube marked 10%.

$TCID_{50}$ Assay

You will use two 96-well plates. The first plate will be used by you to prepare the $TCID_{50}$ dilutions. The second plate will be prepared for you and will contain approximately 1×10^4 cells/well and 150 µl growth medium in each well. The dilutions prepared in the first plate will be transferred to the second plate in the step 4.

Step 3 TCID Dilutions

Take both plates and place and mark them out as depicted in the sketch below. Fill all wells of plate 1 with 180 µl of medium using a multipette.

To start your dilution series from H to A ($1:10$ to $1:10^8$), in each line start by adding 20 µl of the respective inactivated mix to the first well (well H) of the dilution series. Deposit the tip, take a new tip, mix the dilution gently by agitating with the pipette, and transfer 20 µl of the first dilution mixture to the next tube. Continue the dilution series down to well A using a new tip for each step. Do this for all your inactivated mixes and the toxicity control.

Step 4 TCID Assay

Take the multipette and transfer 50 µl from all the wells of each line of your prepared dilutions on the first plate to the respective line of the second plate. This will result in dilutions from $1:4 \times 10^1$ to $1:4 \times 10^8$.

Cover the plate with the lid and place the plate into the incubator at 37 °C/5% CO_2 overnight.

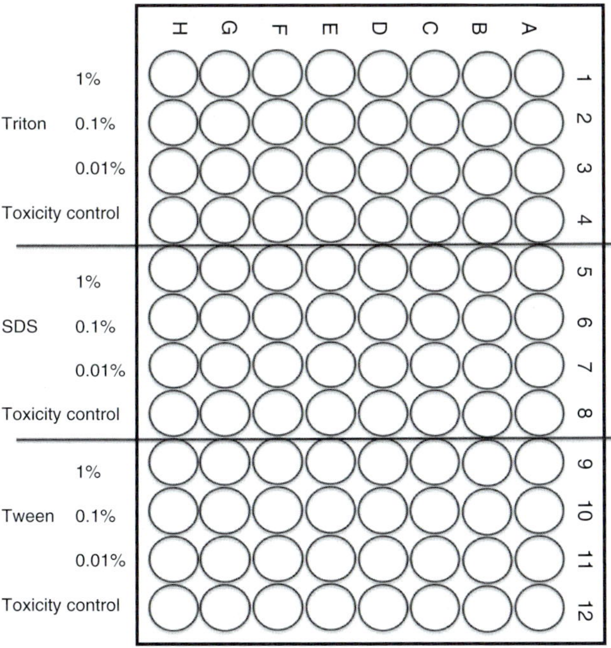

Outline of the $TCID_{50}$ plate. Please mark the first and the second plates as indicated here. The first plate will be used for preparing dilutions. The second plate will contain the cells onto which the dilutions of plate one are transferred and in which the read out will take place on day 3.

Bacillus anthracis Culture

1) Culturing *B. anthracis*
   Read plates.
2) Preparation for polymerase chain reaction (PCR): filtering out spores of *B. anthracis* DNA
   B. anthracis is a spore-forming bacterium. As any material leaving the laboratory should be free of bacteria and spores, we will ask you to purify a *B. anthracis* DNA preparation by using a syringe top filter with 0.2 µm pores.hb
   DNA preparation (already made for you):
   Before taking DNA out of the BSL-3 facility: filtrate through a 0.2 µm, ø 25 mm filter (Millex, Syringe Driven Filter unit, Millipore) into a new microtube.
   To do so:

   a. take out the plunger of the syringe;
   b. screw the 0.2 µM filter disk to the syringe top and arrange this on top of a 2 ml tube;
   c. fill the fluid into the syringe from the back end; and
   d. gently stick back the plunger and slowly press it down.

   Take a loop of the purified DNA and spread it onto a blood agar plate and place into the incubator at 37 °C over night.
   The DNA can only be taken out of the laboratory if the plate is negative on the next day.
3) Decontamination control
   Decontaminate the laboratory bench by wiping the bench with 2%, 5% Wofasteril + Alcapur combination procedure.
   Mix: 900 ml water + 75 ml Alcapur + 25 ml Wofasteril.
   Clean the pipettes and other equipment within the bench.
   Put cellulose on the bench surface and pour the disinfectant on the cellulose.
   Let the disinfectant act for 15 min. Remove the wet cellulose from the bench and clean the bench.
   Use surface sampling agar plates to test for remaining spore activity and place into the incubator over night.

Day 3

Dexterity – Training

Working in a BSL-3 laboratory all dressed up and wearing three gloves requires a lot of preparation and concentration. To accustom you to the situation, a mock pipetting scheme mimicking a typical dilution series after a virus culture has to be done twice once with colored solutions and once with clear solutions. This mock dilution series is pipetted as "blue-in-blue solution" in the first round and "clear-in-clear solution" in the second round.

1) Preparing supernatant stocks
   In a 50 ml tube, you will find 10 ml of mock virus culture supernatant. Pipette 5 × 1 ml into cryotubes and close the lids.
2) Dilution series
   To prepare the dilution series, add 900 µl of the "dilution buffer" into five 1.5 ml tubes.
   Start the dilution series by adding 100 µl of the "virus supernatant" to the first tube of the dilution series. Deposit the tip, take a new tip, mix the dilution gently by agitating with the pipette, and transfer 100 µl of the first dilution mixture to the next tube. Continue the dilution series down to tube five using a new tip for each step.

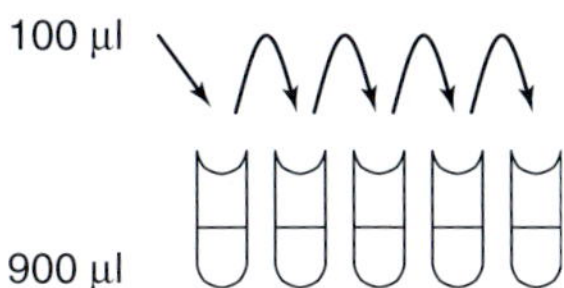

3) Inoculation of six-well plate
   Transfer 500 µl of the dilution buffer into the first six-well plate cavity (negative control). Transfer 500 µl of each dilution into a separate six-well plate cavity.

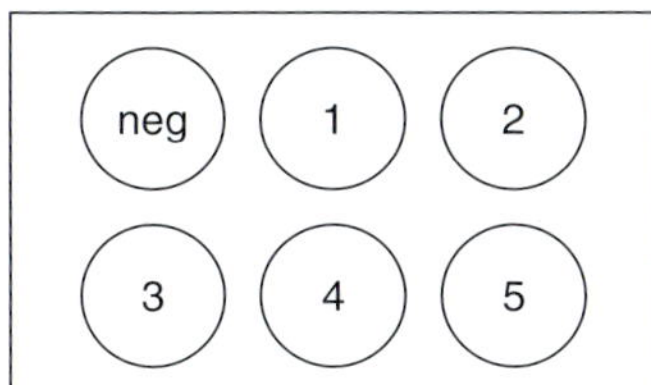

Repeat this protocol using the colorless reagents.

Virus Inactivation

- Take the 96-well plates out of the incubator. Place them under the microscope and look for the well, which show cell lysis.
- Use plate sketch that is shown in *Outline of the $TCID_{50}$ plate* to mark wells with intact cell cultures with plus and cell cultures that have been lysed by virus growth with minus.
- Please note that in real setup four lines of wells for each parameter and that the experiment needs to be performed in triplicate and $TCID_{50}$ values need to be calculated.

Bacillus anthracis Culture

Read plates.

Please note that they need to be negative before the DNA tubes can be externally inactivated by inactivation agent and taken out of the BSL-3 containment.

Index

Working in Biosafety Level 3 and 4 Laboratories: A Practical Introduction, First Edition.
Edited by Manfred Weidmann, Nigel Silman, Patrick Butaye, and Mandy Elschner.